CBS PROBLEMS and
SOLUTIONS *Series*

Problems and Solutions in
POWER
ELECTRONICS

- 📖 This book is suitable for examinations of all Indian universities.
- 📖 Solutions of **150 problems** are given in most simplified manner.
- 📖 All types of problems are included and the solutions given in detail.
- 📖 To explain the basics lucidly, computerised figures have been liberally added in the solutions. This volume contains more than **225 line diagrams**.

CBS Problems and Solutions *Series*

- 📖 Engineering Thermodynamics
- 📖 Strength of Materials
- 📖 Integrated Electronics
- 📖 Communication Systems
- 📖 Signals and Systems
- 📖 Power Systems
- 📖 Electrical Machines and Transformers
- 📖 Network Analysis
- 📖 Power Systems
- 📖 Control Systems
- 📖 Analog Systems

plus many ... more new topics

CBS PROBLEMS and SOLUTIONS Series

Problems and Solutions in POWER ELECTRONICS

R. GOPAL
Engineer

CBS

CBS PUBLISHERS & DISTRIBUTORS
New Delhi • Bangalore

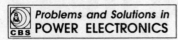

Problems and Solutions in POWER ELECTRONICS

First Edition : 2005
Reprint: 2008

ISBN : 81-239-1235-8

Production Director : Vinod K. Jain

Published by :
Satish Kumar Jain for CBS Publishers & Distributors,
4596/1-A, 11 Darya Ganj, New Delhi - 110 002 (India)
E-mail : cbspubs@del3.vsnl.net.in
Website : http://www.cbspd.com
Seema House, 2975, 17th Cross, K.R. Road,
Bansankari 2nd Stage, Bangalore - 560070
Fax : 080-6771680 • E-mail : cbsbng@vsnl.net

Printed at India Binding House, Noida (UP).

Preface

Power Electronics is one of the essential subjects for the students of electronics, electronics and telecommunication and electrical engineering. The author feels it a great pleasure in presenting this very complex subject through typical problems and their solutions.

This book covers all important and major topics taught at the undergraduate level in an engineering college. All the **150 problems** have been discussed at length, and detailed **225 line diagrams** given extensively to help the reader understand the _basics_ and _grasp the logic_. This book is suitable for preparing for all university examinations and also to face the competitions. This feature is expected to help the reader in visualising the patterns of problems and the extent and level of difficulty in solving the questions likely to be encountered in the examinations, and thus tackling the problems with confidence and success.

Although every care has been taken to ensure accuracy, yet some errors might have crept in. The author will be grateful if these are brought to his notice so that these could be rectified in the subsequent printings and editions of the book. Readers are requested to send their suggestions for further improvement and revision to me through the publishers at the following e-mail address: cbspubs@del3.vsnl.net.in.

R. GOPAL

Contents

CHAPTER 1

DIODE CIRCUITS AND RECTIFIERS

P-N junction diode: Power diode is a two-layer, two terminal, *p-n* semiconductor device. It has one *p-n* junction formed by alloying, diffusing or epitaxial growth. The two terminals of diode are called anode and cathode.

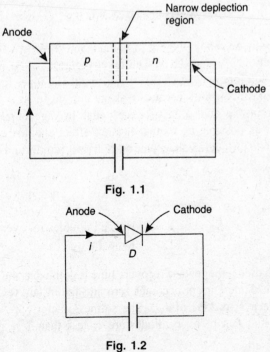

Fig. 1.1

Fig. 1.2

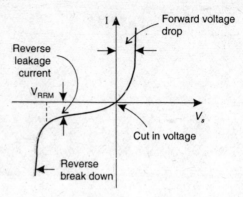

Fig. 1.3

Ge vs Si: (1) Energy gap at 27°C $\qquad \left\{ \begin{array}{l} \text{Ge} : 0.7 \text{ V} \\ \text{Si} : 1.1 \text{ V} \end{array} \right.$

at 0°C $\qquad \left\{ \begin{array}{l} \text{Ge} : 1.1 \text{ V} \\ \text{Si} : 1.21 \text{ V} \end{array} \right.$

(2) Cut in voltage $\qquad \left\{ \begin{array}{l} \text{Ge} : 0.3 \text{ V} \\ \text{Si} : 0.7 \text{ V} \end{array} \right.$

(3) Reversed biased voltage $\left\{ \begin{array}{l} \text{Ge} : 15 \text{ V} \\ \text{Si} : 13 \text{ V} \end{array} \right.$

In a silicon diode, the cut in voltage and forward voltage drop are respectively 0.7 V and 0.6 V. The reverse current of Si is of the order of 10 nA and for Ge is 1 to 2 μA, that is about 100 times less. The device is made of silicon only because leakage current in Si is much smaller than that in Ge and it is necessary that leakage current should be minimum as possible as so that heating effect should be minimum.

Beyond the cut in voltage, the current rises rapidly and exponentially.

$$I_D = I_O \left(e^{V_D/E_T^{\eta}} - 1\right), \qquad \eta = 2 \text{ for low voltage}$$
$$\eta = 1 \text{ for high voltage}$$

$$E_T = \frac{T}{11,000}, \text{ Hence I doubles for every } 10°\text{C rise in temperature.}$$

For a diode, the reverse recovery time is defined as the time between the instant diode current becomes zero and the instant reverse recovery current decays to 25% of I_{RM}. The softness factor S (= t_b/t_a) for soft recovery and fast recovery diodes are 1, less than 1 respectively.

There are three types of power diodes:
(1) **General purpose diodes:** These diode have relatively high reverse recovery time, of the order of about 25 μs. These diodes are basically used in electric traction, electroplating, UPS.
(2) **Fast recovery diodes:** These diodes have relatively low recovery time of about 5 μs or less. These are used in chopper, commutation circuit, induction heating etc.
(3) **Schottky diodes:** Schottky diodes are characterised by very fast recovery time and low forward voltage drop. Their reverse voltage rating are limited to about 100 V and forward current ratings vary from 14 to 300 A.

Power transistors: These are classified into three types:
(1) Bipolar junction transistor (BJT)
(2) Metal-oxide semiconductor field-effect transistor (MOSFET)
(3) Insulated gate bipolar transistor (IGBT)

(1) BJT: A BJT is a three layer, three terminal, two junction *p-n-p* or *n-p-n* semiconductor device.

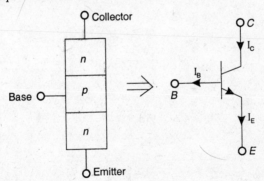

Fig. 1.4

Input characteristics:

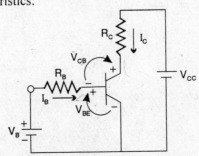

Fig. 1.5

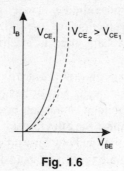

Fig. 1.6

Output characteristics:

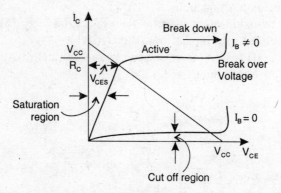

Fig. 1.7

Transistor biasing:

E-B	C-B	Abbre.	Type of operation
Forward (F)	Reverse (R)	F-R	Active region
F	F	F-F	Saturation
R	R	R-R	Cut-off
R	F	R-F	Inverted

Relation between α and β: Forward current gain 'α' is defined as:

$$\alpha = \frac{I_C}{I_E}$$

As $I_C < I_E$, value of a varies from 0.95 to 0.99. Current gain β is defined as

$$\beta = \frac{I_C}{I_B}$$

I_B is much smaller than I_C. So B is much more than unity; its value varies from 50 to 300.

$\therefore$ $\qquad I_E = I_C + I_B$

or, $\qquad \dfrac{I_E}{I_C} = 1 + \dfrac{I_B}{I_C}$

or, $\qquad \dfrac{1}{\alpha} = 1 + \dfrac{1}{\beta}$

or, $\qquad \boxed{\beta = \dfrac{\alpha}{1-\alpha}}$

and $\qquad \boxed{\alpha = \dfrac{\beta}{1+\beta}}$

(2) MOSFET : BJT is a current controlled device where as MOSFET is a voltage controlled device power MOSFET conduction is due to majority carrier, therefore time delays caused by removal or recombination of minority carriers are eliminated. There are two types of MOSFET: (a) Enhancement type MOSFET, (b) Depletion type MOSFET.

(a) **Enhancement Type MOSFET :** It is of two types:

    (1) *n*-channel enhancement MOSFET.

    (2) *p*-channel enhancement MOSFET.

In E-MOSFET there is no channel between the drain and source but the *n*-type drained and the source are separated by a *p*-type substrate. Over the surface of the substrate a very thin Layer of SiO_2 is deposited. A metallic film is deposited on the surface of SiO_2 which acts as the gate. The input impedance of E-MOSFET is very high of the order of $10^{15}\ \Omega$.

The main disadvantage of *n*-channel MOSFET is that conducting *n*-channel in between drain and source gives large on-state resistance. This leads to high power loss in *n*-channel.

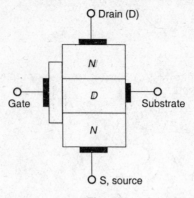

Fig. 1.7

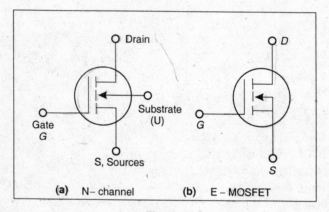

(a) N– channel   **(b)** E – MOSFET

Fig. 1.8

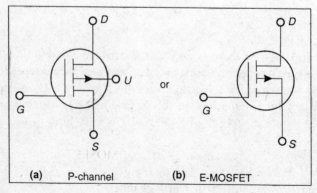

(a) P-channel   **(b)** E-MOSFET

Fig.1.9

(b) **Depletion type MOSFET :** It is similar to E-MOSFET in construction except that a lightly doped *N*-type channel is introduced between the two heavily doped source and drains. It can be operated with +ve and –ve gate while E-MOSFET operated with –ve gate potential only.

(a) *N*-channel DMOSFET.

(b) *p*-channel D-MOSFET.

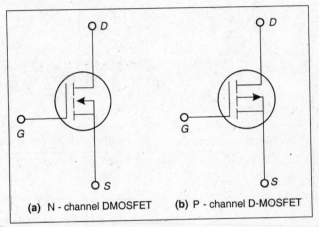

(a) N - channel DMOSFET   (b) P - channel D-MOSFET

Fig. 1.10

Comparision of MOSFET and BJT.

(1) MOSFET is a voltage controlled device where as BJT is a current controlled device.

(2) MOSFET has lower switching losses but its on resistance and conduction losses are more as compare to BJT. So at high frequency power MOSFET is superior and at lower frequency BJT is superior.

(3) MOSFET has +ve temperature co-efficient for resistance but a BJT has –ve temperature co-efficient.

(4) In MOSFET secondary break down does not occures while in BJT secondary break down occure.

(3) IGBT : This device combines into it the advantage of both MOSFET and BJT. So an IGBT has high input impedance like a MOSFET and low on-state power loss as in a BJT.

IGBT is free from secondary break down problem present in BJT. It is widely used in medium power application. Such as d.c. and a.c. motor drives, UPS, relays and contactors etc.

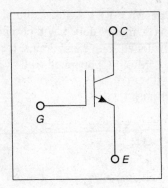

Fig.1.11

Diode rectifications

S.N	Parameters	HW rectifier	FW rectifier
1.	$I_{r.m.s}$	$\dfrac{I_m}{2}$	$\dfrac{I_m}{J_2}$
2.	$I_{d.c.}$	$\dfrac{I_m}{\pi}$	$\dfrac{2I_m}{\pi}$
3.	$V_{d.c.}$	$\dfrac{I_m \cdot R_L}{\pi}$	$\dfrac{2I_m R_L}{\pi}$
4.	% Rectifier efficiency	$\dfrac{40 \cdot 5}{1 + \left(\dfrac{R_f}{R_L}\right)}$	$\dfrac{81.0}{1 + \left(\dfrac{R_f}{R_L}\right)}$
5.	Ripple factor	$1 \cdot 21$	0.48
6.	Fundamental ripple frequency	F	$2F$
7.	Peak Inverse Voltage	V_m	$2V_m$

Zener diodes

Zener diodes are specially constructed to have accurate and stable reverse break down voltage. When it is forward biased it behaves as a normal diode. When reverse biased, a small leakage current flows. If the reverse voltage is increased, a value of voltage will reach at which reverse break down occurs. The voltage after reverse break down remain practically constant over a wide range of zener current.

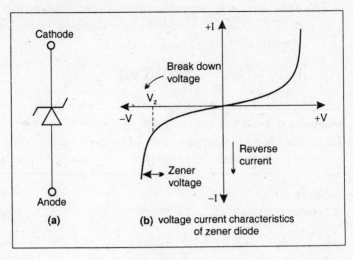

Fig. 1.12

Advantages

Advantage of zener diode regulator : It is a simple circuit, light weight, more reliable and provides regulation over a wide range of current.

Disadvantages:
(1) As there is power dissipation in series resistor and the diode, it results in *poor efficiency.*
(2) The stabilized output is determined by the zener break down voltage and cannot be varied.
(3) Effects of ambient temperature change can cause the break down voltage to change.

Note : The zener break down below 6 V occurs for heavily doped junctions.

Power semiconductor devices

(1) **RCTs and GATTs :** These are widely used for high speed switching, especially intraction application.
(2) **TRIACs :** These are widely used in all types of simple heat controls, light controls, motor controls and a.c switches.
(3) **LASCRs :** These are widely used in high-voltage power systems, especially in HVDC.
(4) **GTOs and SITHs :** These are self commutated thyristors.
(5) **Power MOSFET :** These are used in high-speed power converters.

(6) **SITs :** It is a high power, high frequency device. It is similar to JFET. It has low-noise, low-distortion, high-audio frequency power capability.

PROBLEMS

Problem 1.1: For a power diode, the reverse recovery time is 3.9 μs and the rate of diode-current decay is 50 A/μs. For a softness factor of 0.3, calculate the peak inverse current and storage charge.

Solution:

Given $t_{rr} = 3.9$ μs

Softness factor $(t_b/t_a) = 0.3$ ∴ $\boxed{t_b = 0.3\, t_a}$

$$t_{rr} = t_a + t_b = 3.9 \text{ μs}$$

or $1.3\, t_a = 3.9$ μs

or $\boxed{t_a = 3 \text{ μs}}$

given : $\dfrac{di}{dt} = 50$ A/μs

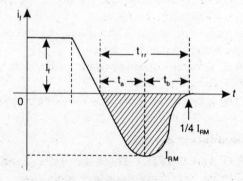

Fig. P. 1.1

The peak inverse current I_{RM} can be expressed as :

$$I_{RM} = t_a \frac{di}{dt} = 3 \times 50 \text{ Amp}$$

or, $\boxed{I_{RM} = 150 \text{ Amp}}$

Storage charge Q_R is given by,

$$Q_R = \frac{1}{2} I_{RM} . t_{rr} = \frac{1}{2} \times 150 \times 3.9 \ \mu C$$

$$\boxed{Q_R = 292.5 \ \mu C}$$

$\therefore$

and

$$\boxed{\begin{array}{l} I_{RM} = 150 \ Amp \\ Q_R = 292.5 \ \mu C \end{array}}$$

Problem 1.2: Switching wave forms of a power transistor is shown in Fig. P. 1.2. If the average power loss in the transistor is limited to 300 W, find the switching frequency at which this transistor can be operated.

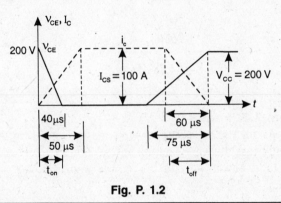

Fig. P. 1.2

Solution: Given

$$I_{CS} = 100 \ A$$
$$v_{CE} = 200 \ V$$
$$t_{on} = 40 \ \mu s$$
$$t_{off} = 60 \ \mu s$$

Energy loss during turn-on period :

$$= \int_0^{t_{on}} i_c \cdot v_{CE} \ dt = \int_0^{t_{on}} \left(\frac{I_{CS}}{50} \times 10^6 t \right) \left(V_{CC} - \frac{V_{CC}}{40} \times 10^6 t \right) dt$$

$$= \int_0^{t_{on}} (2 \times 10^6 t)(200 - 5 \times 10^6 t) \ dt = 0.1066 \ \text{Watt-Sec}$$

Energy loss during turn off period :

$$= \int_0^{t_{\text{off}}} \left(I_{CS} - \frac{I_{CS}}{t_{\text{off}}} t \right) \left(\frac{V_{CC}}{75} \times 10^6 t \right) dt$$

$$= \int_0^{t_{\text{off}}} \left(100 - \frac{100}{60} \times 10^6 t \right) \left(\frac{200}{75} \times 10^6 t \right) dt$$

$$= 0.1600 \text{ Watt-Sec}$$

$\therefore$ Total energy loss during one cycle

$$= (0.1066 + 0.1600) \text{ W-Sec} = 0.2666 \text{ Watt-Sec}$$

Average power loss in transistor

$$= \text{Switching frequency} \times \text{energy loss in one cycle}$$
$$300 = f \times 0.2666$$

or $$f = \frac{300}{0.2666} \text{ Hz}$$

$$\boxed{f = 1125.28 \text{ Hz}}$$

Problem 1.3: In case $I_{CS} = 80$ A, $V_{CC} = 200$ V, $t_{\text{on}} = 1.5$ μs and $t_{\text{off}} = 4$ μs for the switching wave forms shown in Fig. P. 1.3, find the energy loss during switch-on and switch-off intervals. Find also the average power loss in the power transistor for a switching frequency of 2 kHz.

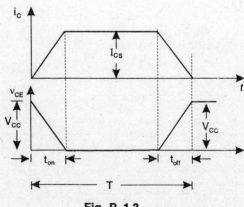

Fig. P. 1.3

Solution: Energy loss during switch on interval is given as :

$$= \frac{V_{CC} \cdot I_{CS}}{6} t_{on} = \frac{220 \times 80}{6} \times 1.5 \times 10^{-6} \text{ Watt-Sec}$$

$$= 4.4 \text{ mWs}$$

Energy loss during switch-off interval is given as :

$$= \frac{V_{CC} \cdot I_{CS}}{6} t_{off} = \frac{220 \times 80}{6} \times 4 \times 10^{-6} \text{ Watt-Sec}$$

$$= 11.73 \text{ mWs}$$

Average power loss in the power transistor for a switching frequency of 2 kHz is given as :

$$= \frac{V_{CC} \cdot I_{CS}}{6} f \cdot t_{on} + \frac{V_{CC} \cdot I_{CS}}{6} f \cdot t_{off}$$

$$= \left(\frac{220 \times 80}{6} \times 2 \times 10^3 \times 1.5 \times 10^{-6} + \frac{220 \times 80}{6} 2 \times 10^3 \times 4 \times 10^{-6} \right) W$$

$$= (8.8 + 23.46) \text{ W} = 32.26 \text{ W}$$

Problem 1.4: (a) For the typical switching wave forms shown in Fig. P. 1.3 for a power transistor, find expression that give peak instantaneous power loss during t_{on} and t_{off} interval respectively.

(b) In case $I_{CS} = 80$ A, $V_{CC} = 220$ V, $t_{on} = 1.5$ μs and $t_{off} = 4$ μs, find the peak value of instantaneous power loss during t_{on} and t_{off} intervals respectively.

Solution: (a) Instantaneous power loss during rise time is given as :

$$P_r(t) = \frac{I_{CS}}{t_r} \cdot t \left\{ V_{CC} - \frac{V_{CC}}{t_r} \cdot t \right\} \qquad ...(1)$$

$\dfrac{dP_r(t)}{dt} = 0$ gives time t_m at which instantaneous power loss during t_r would be maximum. So,

$$\frac{dP_r(t)}{dt} = \frac{I_{CS} V_{CC}}{t_r} \left(1 - \frac{2t_m}{t_r} \right) = 0$$

or $\qquad 1 - \dfrac{2t_m}{t_r} = 0$

or $\qquad t_m = \dfrac{t_r}{2}$

∴ Peak instantaneous power loss P_{rm} during rise time is obtained by substituting the value of $t = \dfrac{t_r}{2}$ in eq. (1)

∴ $\qquad P_{rm} = \dfrac{I_{CS}}{t_r} \cdot \dfrac{t_r}{2}\left(V_{CC} - \dfrac{V_{CC}}{t_r}\cdot\dfrac{t_r}{2}\right) = \dfrac{I_{CS}\cdot V_{CC}}{2}\left(1 - \dfrac{1}{2}\right)$

$$\boxed{P_{rm} = \dfrac{I_{CS}\cdot V_{CC}}{4}}$$

Instantaneous power loss during fall-time is

$$P_f(t) = I_{CS}\left[1 - \dfrac{t}{t_f}\right]\left[\dfrac{V_{CC}}{t_f}\times t\right] \qquad\qquad ...(2)$$

$\dfrac{d\,P_t(t)}{dt} = 0$ gives time t_m at which instantaneous power loss during t_r would be maximum.

Hence $\qquad t_m = \dfrac{t_f}{2}$

Peak instantaneous power loss Pf_m during fall time is obtained by substituting the value of $t = \dfrac{t_f}{2}$ in eq. (2) So,

$$P_{fm} = I_{CS} = \left[1 - \dfrac{t_{f/2}}{t_f}\right]\left[\dfrac{V_{CC}}{t_f}\cdot\dfrac{t_f}{2}\right]$$

$$\boxed{P_{fm} = \dfrac{I_{CS}\cdot V_{CC}}{4}}$$

(b) For $I_{CS} = 80$ A and $V_{CC} = 220$ V, $t_{on} = 1.5$ μs, $t_{off} = 4$μs. The peak instantaneous power loss during t_{on} intervals is given as:

$$= \dfrac{I_{CS}\cdot V_{CC}}{4} = \dfrac{80\times 220}{4}\ \text{W} = 4400\ \text{W}$$

The peak instantaneous power loss during t_{off} intervals is given as:

$$= \frac{I_{CS} \cdot V_{CC}}{4} = \frac{80 \times 220}{4} \text{ W} = 4400 \text{ W}$$

Problem 1.5 : 4 power transistor is used as a switch and typical wave forms are shown in Fig. P. 1.4. The parameters for the transistor circuit are as under :

V_{CC} = 200 V, V_{CES} = 2.5 V, I_{CS} = 60 A, t_d = 0.5 μs, t_r = 1 μs, t_n = 40 μs, t_s = 4 μs, t_f = 3 μs, t_0 = 30 μs, f = 10 kHz

Collector to emitter leakage current = 1.5 mA. Determine average power loss due to collector current during t_{on} and t_n. Find also the peak instantaneous power loss due to collector current during turn-on time.

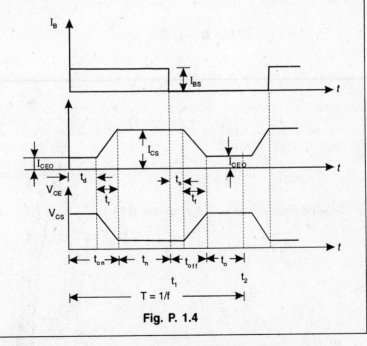

Fig. P. 1.4

Solution: Instantaneous power loss during delay time is

$$P_d (t) = i_C \cdot V_{CE} = I_{CEO} \cdot V_{CC}$$
$$= 1.5 \times 10^{-3} \times 200 = 0.30 \text{ W}$$

Average power loss during delay time is given by

$$P_d = \frac{1}{T}\int_0^{t_d} i_C(t) \cdot V_{CC}(t)\, dt$$

or $\quad P_d = \frac{1}{T}\int_0^{t_d} I_{CEO} \cdot V_{CC}\, dt = \frac{1}{T} I_{CEO} \cdot V_{CC} \cdot t_d$

$$= 10 \times 10^3 \times 1.5 \times 10^{-3} \times 200 \times 0.5 \times 10^{-6}$$
$$= 1.5 \text{ mW}$$

During rise time $0 \leq t \leq t_r$

$$i_C(t) = \frac{I_{CS}}{t_r} \cdot t$$

and $\quad V_{CE}(t) = \left[V_{CC} - \frac{V_{CC} - V_{CES}}{t_r} \cdot t \right]$

$\therefore$ Average power loss during rise time is

$$P_r = \frac{1}{T}\int_0^{t_r} \frac{I_{CS}}{t_r} \cdot t \left[V_{CC} - \frac{V_{CC} - V_{CES}}{t_r} \cdot t \right] dt$$

$$= \frac{1}{T} I_{CS} \cdot t_r \left[\frac{V_{CC}}{2} - \frac{V_{CC} - V_{CES}}{3} \right]$$

$$= 10 \times 10^3 \times 60 \times 10^{-6} \left[\frac{200}{2} - \frac{200 - 2.5}{3} \right] = 20.50 \text{ W}$$

Intantaneous power loss during rise time is

$$P_r(t) = \frac{I_{CS}}{t_r} \cdot t \left\{ V_{CC} - \frac{V_{CC} - V_{CES}}{t_r} \cdot t \right\}$$

$$= \frac{I_{CS} \cdot t}{t_r} V_{CC} - \frac{I_{CS} \cdot t^2}{t_r^2} \left[V_{CC} - V_{CES} \right] \qquad \qquad ...(1)$$

$\dfrac{dP_r(t)}{dt} = 0$ gives time t_m at which intantaneous power loss during t_r would be maximum.

$$\therefore \qquad t_m = \frac{V_{CC} \cdot t_r}{2[V_{CC} - V_{CES}]} = \frac{200 \times 10^{-6}}{2[200 - 2.5]}$$

$$\boxed{t_m = 0.50 \ \mu s}$$

Peak instantaneous power loss P_m during rise time is obtained by substituting the value of $t = t_m$ in eq. (1)

$$\therefore \qquad P_m = \frac{I_{CS} \cdot V_{CC}^2}{4[V_{CC} - V_{CES}]} = \frac{60 \times (200)^2}{4[200 - 2.5]}$$

$$\boxed{P_m = 3038 \ W}$$

Total average power loss during turn on
$$P_{on} = P_d + P_r = (1.5 \times 10^{-3} + 20.50) \ W$$

$$\boxed{P_{on} = 20.5015 \ W}$$

During conduction time, $0 \le t \le t_n$
$$i_C(t) = I_{CS} \text{ and } V_{CE}(t) = V_{CES}$$
$\therefore$ Instantaneous power loss during t_n is
$$P_n(t) = i_C(t) \cdot V_{CE}(t) = I_{CS} \cdot V_{CES}$$
$$= 60 \times 2.5 = 150 \ W$$
Average power loss during conduction period is

$$P_n = \frac{1}{T} \int_0^{t_n} I_{CS} V_{CES} \, dt = \frac{1}{T} I_{CS} \cdot V_{CES} t_n$$

$$= 10 \times 10^3 \times 60 \times 2.5 \times 40 \times 10^{-6}$$

$$\boxed{P_n = 60 \ W}$$

Problem 1.6 : Repeat Prob. P. 1.5 for obtaining average power loss during turn-off time and off peirod and also peak instantaneous power loss during turn-off time due to collector current. Sketch the instantaneous power loss during turn-off time and off-period.

Solution: During storage time $0 \le t \le t_s$
$$i_C(t) = I_{CS} \text{ and } V_{CE}(t) = V_{CES}$$

∴ Instantaneous power loss during t_s is

$$P_s(t) = i_C(t) \cdot V_{CE}(t) = I_{CS} \cdot V_{CES} = 60 \times 2.5$$

$$\boxed{P_s(t) = 150 \text{ W}}$$

Average power loss during t_s is

$$P_s = \frac{1}{T}\int_0^{t_c} I_{CS} \cdot V_{CE} \, dt = f \times I_{CE} \times V_{CES} \times t_s$$

$$= 10 \times 10^3 \times 60 \times 2.5 \times 4 \times 10^{-6}$$

$$\boxed{P_s = 6 \text{ W}}$$

During fall time $0 \le t \le t_f$

$$i_C(t) = \left[I_{CS} - \frac{I_{CS} - I_{CEO}}{t_f} \cdot t \right]$$

During t_f, I_{CEO} is negligibly small as compare to I_{CS}.

$$\therefore \qquad i_C(t) = I_{CS}\left[1 - \frac{t}{t_f} \right]$$

and $\qquad V_{CE}(t) = \dfrac{V_{CC} - V_{CES}}{t_f} \cdot t$

∴ Average power loss during fall time t_f is

$$P_f = \frac{1}{T}\int_0^{t_f} I_{CS}\left(1 - \frac{t}{t_s}\right)\left[\frac{V_{CC} - V_{CES}}{t_f} \cdot t\right] dt$$

$$= f(V_{CC} - V_{CES}) \cdot t_f \left[\frac{I_{CS}}{2} - \frac{I_{CS}}{3}\right]$$

$$= f \cdot t_f \frac{I_{CS}}{6}[V_{CC} - V_{CES}]$$

$$= 10 \times 10^3 \times 3 \times 10^{-6} \times \frac{60}{6}[200 - 2.5] = 59.25 \text{ W}$$

Instantaneous power loss during fall time is

$$P_f(t) = I_{CS}\left[1 - \frac{t}{t_f}\right]\left[\frac{V_{CC} - V_{CES}}{t_f} \cdot t\right]$$

$\dfrac{dP_f(t)}{dt} = 0$ gives time t_m at which instantaneous power loss during t_f, would be maximum.

Here $\quad t_m = \dfrac{t_f}{2}$

∴ Peak instantaneous power loss during t_f is :

$$P_{fm} = I_{CS}\left(1 - \frac{1}{2}\right)\left(\frac{V_{CC} - V_{CES}}{2}\right) = \frac{I_{CS}(V_{CC} - V_{CES})}{4}$$

$$= \frac{60\,(200 - 2.5)}{4} = 2962.5 \text{ W}$$

∴ Total average power loss during turn off procedure is
$$P_{\text{off}} = P_s + P_f = 6 \text{ W} + 59.25 \text{ W}$$

$$\boxed{P_{\text{off}} = 65.25 \text{ W}}$$

During off period $0 \le t \le t_o$
$$i_C(t) = I_{CEO} \text{ and } V_{CE}(t) = V_{CC}$$

Instantaneous power loss during t_o is
$$P_o(t) = i_C \cdot V_{CE} = I_{CEO} \times V_{CC} = 1.5 \times 10^{-3} \times 200 = 0.3 \text{ W}$$

Average power loss during t_o is

$$P_o = \frac{1}{T}\int_0^{t_o} P_o(t)\,dt = f \cdot I_{CEO} \cdot V_{CC} \cdot t_o$$

$$= 10 \times 10^3 \times 1.5 \times 10^{-3} \times 200 \times 30 \times 10^{-6} = 0.090 \text{ W}$$

∴ Total average power loss in power transistor due to collector current over a period T is
$$P_T = P_{\text{on}} + P_n + P_{\text{off}} + P_o$$
$$= (20.5015 + 60 + 65.25 + 0.090) \text{ W} = 145.84 \text{ W}$$

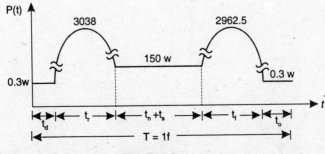

Fig. P. 1.6

Problem 1.7: Capacitor in the circuit of Fig. P. 1.7 is initially charged with (a) V_O volts, (b) $-V_O$ volts. For both these parts, determine the expressions for current in the circuit and voltage across capacitor. Sketech the wave forms for current as well as capacitor voltage. What is the final value of voltage across capacitor in each case?

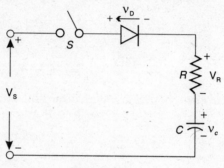

Fig. P. 1.7

Solution: (a) $i(o) = \dfrac{V_S - V_O}{R}$, $i(\infty) = 0$

$\therefore \qquad i(t) = \left(i(o) - i(\infty)\right) e^{-\frac{t}{R_C}} + i(\infty)$

$$i(t) = \left(\frac{V_S - V_O}{R}\right) e^{-\frac{t}{RC}}$$

Voltage across capacitor is

$$V_C(t) = \frac{1}{C}\int_0^t i\, dt + V_O = \left(\frac{V_S - V_O}{RC}\right)\left[\frac{e^{-\frac{t}{RC}}}{-\frac{1}{RC}}\right]_o^t + V_O$$

$$= (V_O - V_S)\left[e^{-\frac{t}{RC}} - 1\right] + V_O$$

$$V_C(t) = (V_S - V_O)\left(1 - e^{-t/RC}\right) + V_O$$

At infinity or in steady-state condition capacitor acts as open circuit. So voltage across capacitor is $= V_S$.

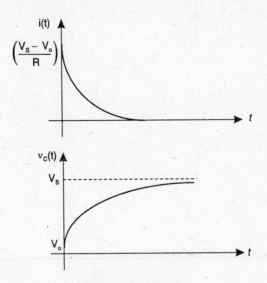

Fig. 1.7 (b) and (c)

(b) When initial voltage across capacitor is $-V_O$ volt.

$$i(o) = \frac{V_S + V_O}{R}$$

$$\therefore \quad i_C(t) = \frac{V_S + V_O}{R} e^{-t/RC}$$

Voltage across capacitor

$$V_C(t) = -(V_S + V_O) e^{-t/RC} + V_S$$
$$= -(V_S + V_O) e^{-t/RC} + (V_S + V_O) - V_O$$

$$V_C(t) = -V_O + (V_S + V_O)\left(1 - e^{-t/RC}\right)$$

At infinity or in steady-state condition capacitor acts as open circuit. So voltage across capacitor is $= V_S$.

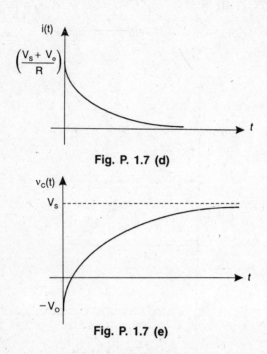

Fig. P. 1.7 (d)

Fig. P. 1.7 (e)

Problem 1.8: For the circuit shown in Fig. P. 1.8, the capacitor is initially charged to a voltage V_O with upper plate positive. Switch S is closed at $t = 0$. Derive expressions for the current in the circuit and voltage across capacitor C. What is the peak value of diode current? Find also the energy dissipated in the ckt.

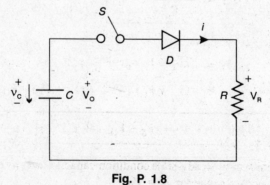

Fig. P. 1.8

Solution: At time $t = 0$, $i(o) = \dfrac{V_O}{R}$

At time $t = \infty$, $i(o) = 0$ (Capacitor is open)

$$\therefore \qquad i(t) = \left(\dfrac{V_O}{R} - 0\right) e^{-t/RC} + 0 = \dfrac{V_O}{R} e^{-t/RC}$$

$\therefore$ Peak diode current $= \dfrac{V_O}{R}$

Capacitor Voltage

$$V_C(t) = \dfrac{1}{C}\int_0^t i\,dt - V_O = \dfrac{1}{C}\int_0^t \dfrac{V_O}{R} e^{-t/RC}\,dt - V_O$$

$$\boxed{V_C(t) = -V_O\, e^{-t/RC}}$$

Energy dissipated in the circuit $= \dfrac{1}{2} C V_O^2$ joules.

Problem 1.9: A diode is connected in series with LC ckt. If this circuit is switched onto d.c. source of voltage V_S at $t = 0$, derive expressions for current through and voltage across capacitor. The capacitor is initially charged to a voltage of $-V_O$. Sketch wave form for i, V_C, V_L and V_D. In case this circuit has $V_S = 230$ V, $V_O = 50$ V, $L = 0.2$ mH and $C = 10$ μF, determine the diode conduction time, diode peak current and final steady state voltage across the capacitor and diode.

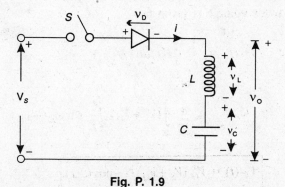

Fig. P. 1.9

Solution: When switch S is closed at $t = 0$,

$$L \frac{di}{dt} + \frac{1}{C} \int i \, dt = V_S$$

Its Laplace transform is

$$L \left[SI(s) - i(o) \right] + \frac{1}{C} \left[\frac{I(s)}{S} + \frac{q(o)}{S} \right] = \frac{V_S}{S}$$

Since $i(o) = 0$, $q(o) = CV_C(o) = -CV_O$.

$$\therefore \qquad I(s) \left[SL + \frac{1}{SC} \right] = \frac{V_S}{S} + \frac{V_O}{S}$$

or $\qquad I(s) = \dfrac{V_S + V_O}{S \left[SL + \dfrac{1}{SC} \right]} = \dfrac{V_S + V_O}{L \left[S^2 + \dfrac{1}{LC} \right]}$

Let $\omega_O = \dfrac{1}{\sqrt{LC}}$, this gives

$$I(s) = \frac{(V_S + V_O)}{L \omega_O} \frac{\omega_O}{S^2 + \omega_O^2}$$

or $\qquad I(s) = (V_S + V_O) \sqrt{\dfrac{C}{L}} \dfrac{\omega_O}{S^2 + \omega_O^2}$

$$\therefore \qquad \boxed{ I(t) = (V_S + V_O) \sqrt{\frac{C}{L}} \sin \omega_O t }$$

Capacitor voltage is given by

$$V_C(t) = \frac{1}{C} \int_0^t i(t) \, dt + (-V_O)$$

$$= \frac{1}{C} \int_0^t (V_S + V_O) \sqrt{\frac{C}{L}} \sin \omega_O t - V_O$$

or $\qquad \boxed{ V_C(t) = V_O^t (V_S + V_O)(1 - \cos \omega_O t) }$

Voltage across inductor is given by

$$V_L(t) = L \frac{di(t)}{dt}$$

$$\boxed{V_L(t) = (V_S + V_O) \cos \omega_O t}$$

Diode voltage after it stops conducting

$$V_D = -V_L - V_C + V_S = 0 - (2 V_S + V_O) + V_S = -(V_S + V_O)$$

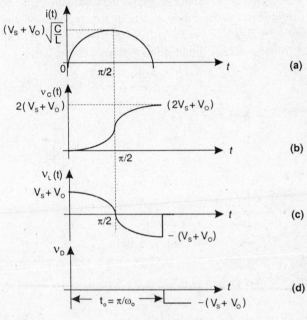

Fig. 1.9 (a) – (d)

$V_S = 230$ V, $V_O = 50$ V, $L = 0.2$ mH, $C = 10.4$ μF

Diode conduction time

$$t_O = \frac{\pi}{\omega_o} = \pi\sqrt{LC} = \pi \sqrt{0.2 \times 10^{-3} \times 10 \times 10^{-6}} = 1.40 \times 10^{-4} \text{ sec}$$

$$\boxed{t_O = 140.49 \text{ μ sec}}$$

Peak current through diode

$$I_P = (V_S + V_O)\sqrt{\frac{C}{L}} = (230 + 50)\sqrt{\frac{10 \times 10^{-6}}{0.2 \times 10^{-3}}}$$

$$\boxed{I_P = 62.61 \text{ Amp}}$$

Stead state voltage across the capacitor is
$$= 2V_S + V_O = 2 \times 230 + 50 = 510 \text{ V}$$
Steady state voltage across diode
$$= -(V_S + V_O) = -(230 + 50) = -280 \text{ V}$$

Problem 1.10 : In the diode and LC network shown in Fig. P. 1.10, the capacitor is initially charged to voltage V_O with upper plate positive. Switch S is closed at $t = 0$. Find the conduction time of diode, peak current through the diode and final steady state voltage across C in case $V_S = 200$ V, $V_O = 50$ V, $L = 100$ µH, $C = 20$ µF. Determine also the voltage across diode after it stops conduction.

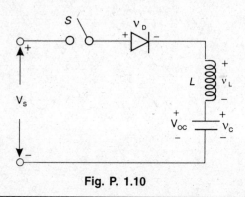

Fig. P. 1.10

Solution: When switch S is closed, KVL gives

$$L\frac{di}{dt} + \frac{1}{C}\int i\,dt = V_S$$

Its laplace transform is given as

$$L\left[SI(s) - i(o)\right] + \frac{1}{C}\left[\frac{I(s)}{S} + \frac{q(o)}{S}\right] = \frac{V_S}{S}$$

Since $i(o) = 0$ and $q(o) = CV_O$

$$\therefore \qquad I(s)\left[SL + \frac{1}{SC}\right] = \frac{V_S - V_O}{S}$$

or $\qquad i(t) = (V_S - V_O) \cdot \sqrt{\frac{C}{L}} \sin \omega_O t$

$$V_C(t) = \frac{1}{C} \int_0^t i(t)\,dt + V_O = (V_S - V_O)(1 - \cos \omega_O t) + V_O$$

4t $\omega_O t = 0$, $V_C(t) = V_O$

and at $\omega_O t = \frac{\pi}{2}$, $V_C(t) = V_S$

and at $\omega_O t = \pi$, $V_C(t) = 2(V_S - V_O) + V_O = 2\ V_sV_O$

Diode conduction time is given as:

$$t_O = \frac{\pi}{\omega_O} = \pi\sqrt{LC} = \pi\ \sqrt{100 \times 10^{-6} \times 20 \times 10^{-6}}$$

$$\boxed{t_O = 140.49\ \mu\ \text{sec}}$$

Peak diode current is equal to

$$I_P = (V_S - V_O)\ \sqrt{\frac{C}{L}} = (200 - 50)\ \sqrt{\frac{20 \times 10^{-6}}{100 \times 10^{-6}}}$$

$$\boxed{I_P = 67.08\ \text{Amp}}$$

Steady state voltage across capacitor is given as :

$$V_C = 2V_S - V_O = 2 \times 200 - 50 = 350\ \text{V}$$

Voltage across diode, after it stops conducting is

$$V_D = -V_L - V_C + V_S = 0 - (2\ V_S - V_O) + V_S = -V_S + V_O$$

$$= -200 + 50 = -150\ \text{V}$$

∴ $$\boxed{V_D = -150\ \text{V}}$$

Problem 1.11 : (a) For the circuit shown in Fig. P. 1.11, the circuit is initially relaxed. If switch S is closed at $t = 0$, sketch the variations of i, V_L, V_C and V_D as a function of time. Derive the expressions decribing these functions.

(b) For part (a), $V_S = 220$ V, $L = 4$ mH, $C = 5\mu$F. Find the diode conduction time and peak diode current. Determine also V_C, V_L, and V_D after diode stops conducting.

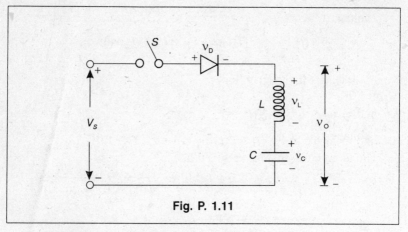

Fig. P. 1.11

Solution: When switch S is closed at $t = 0$,

$$L\frac{di}{dt} + \frac{1}{C}\int i\,dt = V_S$$

Taking laplace transform

$$L\left[SI(s) - i(o)\right] + \frac{1}{C}\left[\frac{I(s)}{S} + \frac{q(o)}{S}\right] = \frac{V_S}{S}$$

Since the circuit is initially relaxed so

$$i(o) = 0 \text{ and } q(o) = 0$$

$\therefore$
$$I(S)\left[SL + \frac{1}{SC}\right] = \frac{V_S}{S}$$

or
$$I(S) = \frac{V_S}{S\left[SL + \dfrac{1}{SC}\right]}$$

Let $\omega_o = \dfrac{1}{\sqrt{LC}}$, this gives

$$I(S) = \frac{V_S}{L\cdot\omega_o}\,\frac{\omega_o}{S^2 + \omega_o^2}$$

$$= V_S\cdot\sqrt{\frac{C}{L}}\,\frac{\omega_o}{S^2 + \omega_o^2}$$

or
$$\boxed{i(t) = V_S\sqrt{\frac{C}{L}}\ \sin\omega_o t}$$

Voltage across capacitor is given as

$$V_C(t) = \frac{1}{C} \int_0^t i(t) \cdot dt = \frac{1}{C} \int_0^t V_S \sqrt{\frac{C}{L}} \sin \varpi_o t \; dt$$

$$\boxed{V_C(t) = V_S(1 - \cos \varpi_o t)}$$

Voltage across inductor

$$V_L(t) = L\frac{di(t)}{dt}$$

or $$\boxed{V_L(t) = V_S \cos \omega_o t}$$

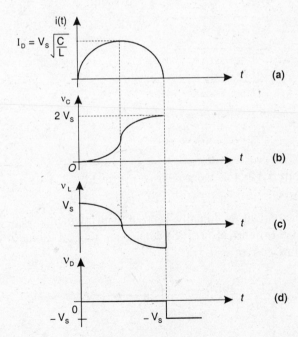

Fig. P. 1.11 (a) – (d)

Voltage across diode soon after diode stops conducting is given by

$$V_D = V_S - V_C - V_L = V_S - 2V_S - 0 = -V_S$$

$$\therefore \qquad \boxed{V_D = -V_S}$$

(b) Diode conduction time is given (a)

$$t_o = \frac{\pi}{\omega_o} = \pi\sqrt{LC} = \pi\sqrt{4\times10^{-3}\times5\times10^{-6}}$$

$$\boxed{t_o = 0.444 \text{ ms}}$$

Peak diode current is given as

$$I_P = (V_S)\sqrt{\frac{C}{L}} = 220\sqrt{\frac{5\times10^{-6}}{4\times10^{-3}}}$$

$$\boxed{I_P = 7.778 \text{ Amp}}$$

After diode stop conducting
$$V_C = 2\,V_S = 2\times220 = 440 \text{ V}$$

$\therefore$

$$\boxed{V_C = 440 \text{ V}}$$

$$\boxed{V_L = 0 \text{ V}}$$

$$\boxed{V_D = -V_S}$$

$$\boxed{V_D = -220 \text{ V}}$$

Problem 1.12: For the circuit shown in Fig. P. 1.12.
$R = 10\ \Omega$, $L = 2$ mH, $C = 6\ \mu$F, $V_S = 220$ V

The circuit is initially relaxed with switch closed at $t = 0$, determine (a) current $i\,(t)$, (b) conduction time of diode, (c) rate of change of current at $t = 0$.

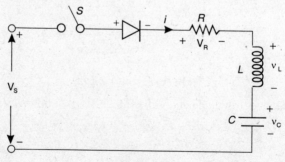

Fig. P. 1.12

Solution: (a) $\xi = \dfrac{R}{2L} = \dfrac{10}{2 \times 2 \times 10^{-3}} = 2500$ rad/sec

$$\omega_0 = \frac{1}{\sqrt{LC}} = \frac{1}{\sqrt{2 \times 10^{-3} \times 6 \times 10^{-6}}} = 91287.09 \text{ rad/s}$$

$$\omega_r = \sqrt{\omega_0^2 - \xi^2} = 91252.85 \text{ rad/sec}$$

Here as $\xi < \omega_0$, the circuit is underdamped.

$$i(t) = \frac{V_S}{\omega_r L} \cdot e^{-\xi t} \sin \omega_r t = \frac{200}{91252.85 \times 2 \times 10^{-3}} e^{-2500 t} \sin (91252.85) t$$

$$\boxed{i(t) = 1.2 \; e^{-2500 t} \sin (91252.85) t}$$

(b) Diode stops conducting when $\omega_r t_1 = \pi$

∴Conduction time of diode

$$t_1 = \frac{\pi}{\omega_r} = \frac{\pi}{91252.85} = 34.42 \times 10^{-6} \text{ sec}$$

$$\boxed{t_1 = 34.42 \; \mu \text{ sec}}$$

(c) $$\frac{di}{dt} = \frac{V_S}{\omega_r L} \left[e^{-\xi t} \omega_r \cos \omega_r t - \sin \omega_r t \cdot (-\xi) e^{-\xi t} \right]$$

$$\left. \frac{di}{dt} \right|_{t=0} = \frac{V_S}{L} = \frac{220}{2 \times 10^{-3}} = 110{,}000 \text{ A/s}$$

Problem 1.13: For the circuit shown in Fig. P. 1.13; $V_o = 230$ V, $R = 25 \; \Omega$ and $C = 10 \; \mu f$. If switch S is closed at $t = 0$, determine

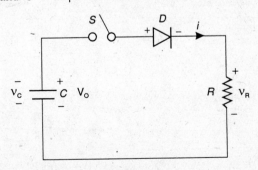

Fig. P. 1.13

expressions for the current in the circuit and voltage across capacitor
C. Find the peak value of diode current and energy lost in the circuit.
Determine the expression used.

Solution: When switch S is closed, KVL gives.

$$R i + \frac{1}{C} \int i\, dt = 0$$

is Taking laplace transform, including the initial voltage across capacitor

$$RI(S) + \frac{1}{C}\left[\frac{I(S)}{S} + \frac{q(0)}{S}\right] = 0$$

or
$$RI(S) + \frac{1}{C}\left[\frac{I(S)}{S} + \frac{CV_o}{S}\right] = 0$$

or
$$I(S)\left[R + \frac{1}{CS}\right] = \frac{V_o}{S}$$

or
$$\boxed{i(t) = \frac{V_o}{R} e^{-t/RC}}$$

Peak diode current $= \dfrac{V_o}{R}$

Capacitor Voltage $V_C(t) = \dfrac{1}{C}\int_0^t i\, dt - V_O = \dfrac{1}{C}\int_0^t \dfrac{V_o}{R} e^{-t/RC}\, dt - V_o$

$$\boxed{V_C(t) = -V_o\, e^{-t/RC}}$$

Energy loss in the circuit $= \dfrac{1}{2} C V_o^2$ Joules

For $V_o = 230$ V, $R = 25\ \Omega$ and $C = 10\ \mu F$

$$i(t) = \frac{230}{25} e^{-t/25} \times 10 \times 10^{-6}$$

$$\boxed{i(t) = 9.2\, e^{-4000t}}$$

$$V_C(t) = -230\ e^{-4000t}$$

Peak diode current $= \dfrac{V_o}{R} = \dfrac{230}{25} = 9.2$ Amp

Energy loss in the circuit $= \dfrac{1}{2} CV_o^2 = \dfrac{1}{2}\ 10 \times 10^{-6}\ (230)^2$

$\qquad\qquad\qquad\qquad = 0.2645$ Watt-Sec

Problem 1.14: In the circuit shown in Fig. P. 1.14 switch's is open and a current of 20A is flowing through the free wheeling diode R and L. If switch S is closed at $t = 0$, determine the expression for the current through the switch.

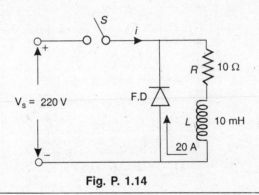

Fig. P. 1.14

Solution:

When switch is open then current through inductor = 20 A

$\qquad \therefore \qquad i(0) = 20$ A

When switch is closed, then free wheeling diode will open circuited.

So,

$$i(\infty) = \dfrac{220}{10} = \dfrac{V_S}{R}$$

$\therefore\ i(\infty) = 22$ Amp

$\therefore$ Expression for the current through the switch is given as.

$$i(t) = \big(i(o) - i(\infty)\big)\ e^{-Rt/L} + i(\infty)$$

$$= (20 - 22) \, e^{-\frac{10t}{10 \times 10^{-3}}} + 22 = -2 \, e^{-1000t} + 22$$

or $$\boxed{i\,(t) \,=\, 22 - 2 \, e^{-1000t}}$$

Problem 1.15 : (a) Describe how the energy trapped in an inductor can be recovered and returned to the source.

(b) A 230 V, 1 kW heater, fed through single-phase half-wave diode rectifier, has rated voltage at its terminals. Find the a.c. input voltage. Find also PIV of diode and peak-diode current.

Solution:

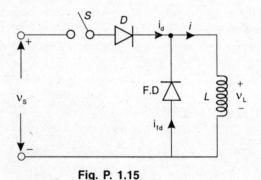

Fig. P. 1.15

In the ideal circuit, of Fig. P.1.15, the energy $\left(\dfrac{1}{2} \left(\dfrac{V_S}{L} t_1 \right)^2 \right)$ stored in

the inductor is trapped. This trapped energy does not dissipated when ever there is no resistance in the circuit. The best way for utilizing this trapped energy is to return it to the source. By doing so, the system efficiency is improved.

One way of returning this trapped energy is to add a second winding closely coupled with the inductor winding as shown in Fig. 1.24 (a).

A diode D is placed in series with the second winding. The inductor now behaves like a transformer. The two winding are so arranged that their polarity marking are opposite to each other.

When switch S is closed, current 'i' begins to flow in the circuit and energy is stored in the inductor of primary winding with N_1 turns. The diode D is reverse biased by voltage $(V_2 + V_S)$.

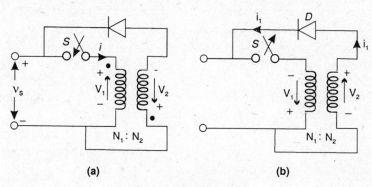

Fig. P. 1.15 (a) and (b)

When switch S is open, the polarities of voltage V_1 and V_2 get reversed, the diode is now forward biased by voltage $(V_2 - V_S)$. As a result, diode begins to conduct a current i, into the positive terminal of source voltage V_S and So the trapped energy is fed back to the source.

Energy feedback to d.c. source

$$= V_S \times \text{current } i_l \text{ depends upon } (V_2 - V_S).$$

(b) Heater resistance $R = \dfrac{(230)^2}{1000} = 52.9\,\Omega$

a.c. input voltage $= \sqrt{2} \times 230\ \text{V} = 325.32\ \text{V}$
PIV of diode $= 2 \times 230\ \text{V} = 460\ \text{V}$

Peak diode current $= \dfrac{2\,V_m}{R} = \dfrac{2 \times 230}{52.9} = 8.696\ \text{Amp}$

Problem 1.16: A $1-\phi$ 220 V,1 kW heater is connected across $1-\phi$ 220 V, 50 Hz supply through a diode. Calculate the power delivered to the heater element. Find also the peak diode current and input power factor.

Solution: Heater resistance, $R = \dfrac{220^2}{1000} = 48.4\ \Omega$

$\therefore$ $\boxed{R = 48.4\ \Omega}$

R.M.S. value of output voltage is $V_{or} = \dfrac{\sqrt{2} \times 220}{2} = 155.56$ V

$\therefore$ $\boxed{V_{or} = 155.56 \text{ V}}$

Power absorbed by heater element $= \dfrac{V_{or}^2}{R} = \dfrac{(155.56)^2}{48.4} = 500$ W

Peak value of diode current is given by

$$= \dfrac{\sqrt{2} \times 220}{(220)^2} \times 1000 = \dfrac{\sqrt{2} \times 1000}{220} = 6.42 \text{ Amp}$$

Input power factor $= \dfrac{V_{or}}{V_S} = \dfrac{220}{\sqrt{2} \times 220} = 0.707$ lag

---

Problem 1.17: (a) A $1-\phi$ half-wave uncontrolled rectifier is connected to RL Load. Derive an expression for the load current in terms of V_m, Z, ω etc.

(b) For part (a), $V_S = 230$ V, at 50 Hz, $R = 10\ \Omega$, $L = 5$ mH, extinction angle $= 210^\circ$. Find average values of output voltage and output current.

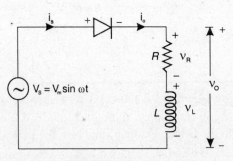

Fig. P. 1.17

Solution: (a) When $i_o = 0$ at $\omega t = \beta$; $V_L = 0$ $V_R = 0$ and voltage V_S appears as reverse bias across diode D as shown. At β, Voltage V_D across diode jumps from zero to $V_m \sin\beta$ where $\beta > \pi$. Here $\beta = y$ is also the conduction angle of the diode.

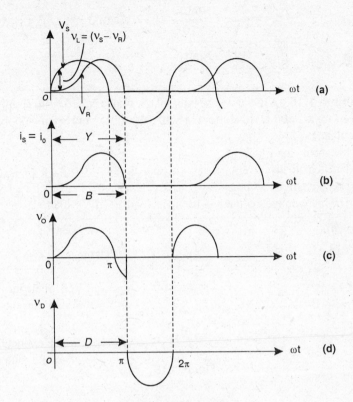

Fig. P. 1.17 (a)-(d)

Average value of output voltage

$$V_o = \frac{1}{2\pi}\int_O^B V_m \sin \omega t.d\,(\omega t) = \frac{V_m}{2\pi}\,(1 - \cos\beta)$$

Average value of load or output current

$$I_o = \frac{V_o}{R} = \frac{V_m}{2\pi R}\,(1 - \cos\,\beta)$$

(b) $\because$ $\quad V_S = V_m \sin \omega t$

$\therefore$ $\qquad \dot{V}_m = \dfrac{230\sqrt{2}}{0.5} = 460\sqrt{2}\ \text{V}$

$\therefore$ $\qquad V_o = \dfrac{460\sqrt{2}}{2\pi}\,(1 - (-0.866))$

$$\boxed{V_o = 193.20\ \text{V}}$$

$$\therefore \qquad I_o = \frac{V_0}{R} = \frac{193.20}{10} = 19.32 \text{ Amp}$$

$$\therefore \qquad \boxed{I_o = 19.32 \text{ Amp}}$$

Problem 1.18 : A diode whose internal resistance is 20 Ω is to supply power to a 1000 Ω load from a 230 V (r.m.s.) source of supply. Calculate :

(a) The peak load current.

(b) The d.c. load current.

(c) The d.c. diode voltage.

(d) The percentage regulation from no load to given load.

Solution:

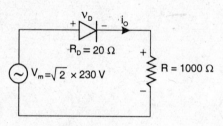

Fig. P. 1.18

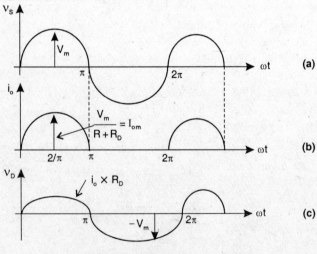

Fig. P. 1.18 (a) - (c)

(a) Peak load current I_{om} is given by (as seen from waveform)

$$I_{om} = \frac{V_m}{R + R_D}$$

$$= \frac{\sqrt{2} \times 230}{1020} = 0.3189 \text{ A}$$

Here R = Load resistance

           R_D = Internal resistance of diode

(b) d.c. load current

$$I_0 = \frac{1}{2\pi} \int_0^{\pi} I_{om} \sin \omega t \, d(\omega t)$$

$$\boxed{I_0 = \frac{I_{om}}{\pi} = 0.1015 \text{ Amp}}$$

(c) d.c. diode voltage

$$V_D = I_o R_D - \frac{1}{2\pi} \int_0^{\pi} 230\sqrt{2} \sin \omega t \, d(\omega t)$$

$$= I_o R_D - \frac{V_m}{\pi} = 0.1015 \times 20 - \frac{230\sqrt{2}}{\pi}$$

$$\boxed{V_D = -101.5 \text{ V}}$$

(d) At no-load, load voltage

$$V_{on} = \frac{V_m}{\pi} = \frac{\sqrt{2} \times 230}{\pi}$$

$$\boxed{V_{on} = 103.521 \text{ V}}$$

At given load, load voltage

$$V_{ol} = \frac{230\sqrt{2}}{\pi} \times \frac{1000}{1020}$$

$$\boxed{V_{ol} = 101.49 \text{ V}}$$

$$\therefore \text{ Voltage regulation} = \frac{V_{on} - V_{01}}{V_{on}} \times 100 = \frac{103.521 - 101.49}{103.521} = 1.96\%$$

V.R. = 1.96%

Problem 1.19: A sinusoidal voltage of amplifier 20 volts and frequency 50 Hz is applied to a half wave rectifier using *p-n* diode. The load resistor is 500 W. Considering idealised characteristic of the diode with $R_f = 5\,\Omega$ and $R_r = \infty$, calculate

(a) peak, average and r.m.s. values of load current.

(b) d.c. power output,

(c) a.c power input,

(d) rectifier efficiency,

(e) ripple factor, and

(f) TUF.

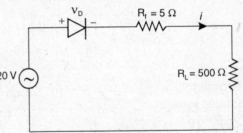

Fig. P. 1.19

Solution: (a) $\quad I_m = \dfrac{V_m}{R_f + R_L} = \dfrac{20}{5 + 500}$

$I_m = 0.0396 \text{ Amp}$

$$I_{d.c.} = \frac{I_m}{\pi} = \frac{39.6 \times 10^{-3}}{2} = 12.6 \text{ mA}$$

$$I_{r.m.s} = \frac{I_m}{2} = \frac{39.6 \times 10^{-3}}{2} = 19.8 \text{ mA}$$

(b) $\quad P_{d.c.} = I_{d.c.}^2 \; R_L = (12.6 \times 10^{-3})^2 \times 500$

$P_{d.c.} = 79.38 \text{ mW}$

(c) $\quad P_{in} = I_{r.m.s.}^2 \; (R_f + R_L) = (19.8 \times 10^{-3})^2 \; (5 + 500)$

$P_{in} = 197.98 \text{ mW}$

(d) Rectifier efficiency $= \dfrac{P_{d.c.}}{P_{a.c.}} = \dfrac{79.38}{197.98} = 40\%$

(e) Ripple factor $y = \sqrt{\left(\dfrac{I_{r.m.s.}}{I_{d.c.}}\right)^2 - 1} = \sqrt{\left(\dfrac{19.8}{12.6}\right)^2 - 1} = 1.21$

(f) $\text{TUF} = \dfrac{0.286}{1 + \dfrac{R_f}{R_L}} = \dfrac{0.286}{1 + \dfrac{5}{500}}$

or $\boxed{\text{TUF} = 0.283}$

Problem 1.20: Design a zener voltage regulator shown in Fig. P.1.20, to meet the following specification. Load voltage = 6.8 V, source voltage V_S is 20 V ± 20% and load current is 20 mA ± 40%. The zener requires a minimum current of 1 mA to breakdown. The diode D has a forward voltage drop of 0.6 V.

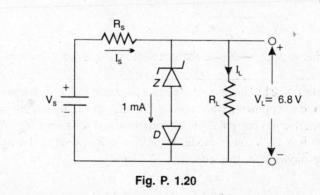

Fig. P. 1.20

Solution: When source voltage is maximum and load current is minimum, then source resistance should be maximum.

$\therefore \qquad V_{S\cdot\max} = V_L + (I_{L\cdot\min} + I_Z)\, R_{S\cdot\max}$

$\therefore \qquad R_{S\cdot\max} = \dfrac{V_{S\cdot\max} - V_L}{(I_{L\cdot\min} + I_Z)} = \dfrac{20 \times 1.2 - 6.8}{(20 \times 0.6 + 1) \times 10^{-3}}$

$$= \frac{17.2}{13 \times 10^{-3}} = 1.32 \times 10^3 \Omega$$

$\therefore$ $\boxed{R_{S \cdot \text{max}} = 1.32 \text{ k}\Omega}$

Similarly,

$$V_{S \cdot \text{min}} = V_L + (I_{L \cdot \text{max}} + I_Z) R_{S \cdot \text{min}}$$

$\therefore$ $R_{S \cdot \text{min}} = \dfrac{V_{S \cdot \text{min}} - V_L}{(I_{L \cdot \text{max}} + I_Z)} = \dfrac{(20 \times 0.8) - 6.8}{[20 \times 1.4 + 1] \times 10^{-3}}$

$\boxed{R_{S \cdot \text{min}} = 317.24 \ \Omega}$

Maximum load resistance

$$R_{L \cdot \text{max}} = \frac{V_L}{I_{L \cdot \text{min}}} = \frac{6.8}{20 \times 0.6 \times 10^{-3}}$$

$\boxed{R_{L \cdot \text{max}} = 566.66 \ \Omega}$

Minimum load resistance

$$R_{L \cdot \text{min}} = \frac{V_L}{I_{L \cdot \text{max}}} = \frac{6.8}{20 \times 1.4 \times 10^{-3}}$$

$\boxed{R_{L \cdot \text{min}} = 242.85 \ \Omega}$

The voltage rating on the zener diode is $= 6.8 - 0.6 = 6.2$ V.

Problem 1.21: The complete circuit shown in Fig. P. 1.21 represents a 25 V d.c. voltmeter where G is a PMMC galvanometer having full-scale deflection current $I_{fsd} = 200$ micro-A and resistance $R_G = 500 \ \Omega$ and D is a 20-V zener diode. Find R_1 and R_2. What is the function of the diode D in this circuit?

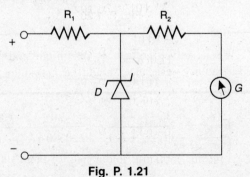

Fig. P. 1.21

Solution: Current through galvanometer

$$I_{fsd} = I_2 = \frac{\text{zener voltage}}{R_2 + R_G}$$

or $\qquad I_2 = \dfrac{20}{R_2 + 500} = 200 \times 10^{-6}$

or $\qquad R_2 = \dfrac{20 \times 10^6 - 500}{200}$

$$\boxed{R_2 = 99.5 \text{ k}\Omega}$$

Let us assumed the zener diode current to be zero.

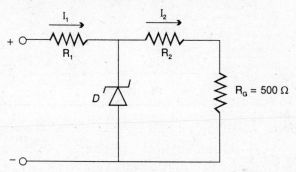

Fig. P. 1.21 (a)

$\therefore \qquad I_1 - I_2 = I_2 = 0$

$\therefore \qquad\qquad I_1 = I_2$

$\therefore \qquad\qquad I_1 = I_2 = 200 \ \mu A$

and $\qquad I_1 = \dfrac{25\,V - 20\,V}{R_1}$

or $\qquad R_1 = \dfrac{5V}{I_1}$

or $\qquad R_1 = \dfrac{5\,V}{200 \times 10^{-6}}$

or $\qquad \boxed{R_1 = 25 \text{ k}\Omega}$

$\therefore \qquad \boxed{R_1 = 25 \text{ k}\Omega \text{ and } R_2 = 99.5 \text{ k}\Omega}$

The function of zener diode is to provide a constant voltage to the circuit. It prevents over loading to the circuit.

Problem 1.22 : For the circuit shown in Fig. P. 1.22, $V_S = 160$ V, $V_Z = 40$ V and zener diode current varies from 4 to 40 mA. Find the minimum and maximum values of R_1 so as to allow voltage regulation for output current $I_O =$ zero to its maximum value I_{om}. Also calculate I_{om}.

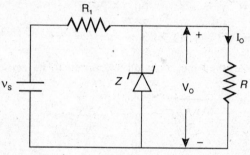

Fig. P. 1.22

Solution: Maximum value of $R_1 = \dfrac{V_S - V_Z}{I_{Z \cdot min}} = \dfrac{160 - 40}{4 \times 10^{-3}}$

for $I_o = 0$

$$= \frac{120}{4 \times 10^{-3}} = 30 \text{ k}\Omega$$

Minimum value of $R_1 = \dfrac{V_S - V_Z}{I_{Z \cdot max}}$

for $I_o = 0$

$$= \frac{160 - 40}{4 \times 10^{-3}} = \frac{120}{4 \times 10^{-3}} = 3 \text{ k}\Omega$$

Maximum value of output current

$$I_{om} = (I_{Z \cdot max} - I_{Z \cdot min}) \text{ according to the question}$$
$$= (40 \text{ mA} - 4 \text{ mA})$$

$$\boxed{I_{om} = 36 \text{ mA}}$$

∴ $\boxed{\begin{array}{l} R_{1 \cdot max} = 30 \text{ k}\Omega, \; R_{1 \cdot min} = 3 \text{ k}\Omega \\ I_{om} = 36 \text{ mA} \end{array}}$

and

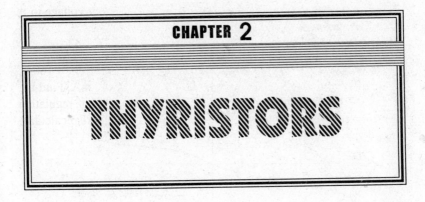

CHAPTER 2

THYRISTORS

Thyristors : It is a four layer, three terminal, three junction semi-conductor device. Like the diode, SCR is an unidirectional device that blocks the current flows from cathode to anode.

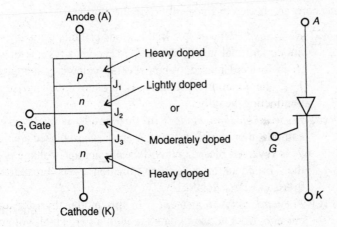

Fig. 2.1 (a) and (b)

Static VI characteristic of a thyristor

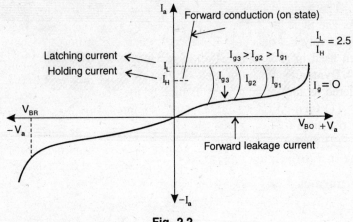

Fig. 2.2

$$V_{BO} = \text{Forward break over voltage}$$
$$V_{BR} = \text{Reverse break over voltage}$$
$$I_g = \text{Gate current}$$

There are basically three mode of the thyristor :

(a) *Reverse blocking mode* : In this mode cathode is made +ve w.r.t. anode terminal junction J_1 and J_3 becomes reverse biased and J_2 is forward biased. When reverse voltage is increases than V_{BR}, the junction J_1 and J_3 is broken and Thyristor starts conducting heavily.

(b) *Forward blocking mode* : In this anode is made +ve w.r.t. cathode, then junction J_1 and J_3 are forward biased and junction J_2 is reversed biased. When forward applied voltage is greater then or equal to V_{BO}, then junction J_2 is breakdown and thyristor starts conducting.

(c) *Forward conduction mode* : In this mode, thyristor conducts currents from anode to cathode with a very small voltage drop across it. In this mode, thyristor is in on-state and behaves like a close switch. All the junction are forward biased.

Thyristor turn-on methods

(a) Forward voltage triggering
(b) Gate triggering
(c) *dv/dt* triggering

(d) Temperature triggering

(e) Light triggering

The typical gate current magnitude are of the order of 20 to 200 mA. The SCR ratings, *di/dt* in A/μ sec and *dv/dt* in V/μ sec, may vary, respectively, between both 20 to 500. Light triggered thyristors have now been used in (HVDC) transmission line.

Dynamic characteristic of thyristors : Dynamic characteristics is reffered as turn-on and turn-off time of the thyristor.

(1) **turn-on time** : It is divided into three parts (a) delay time (t_d) (b) rise time (t_r) (c) spread time (t_p).

(2) **turn-off time** : It is divided into two parts (a) reverse recovery time (t_{rr}) (b) gate recovery time (t_{gr}) or (t_{rc}).

(1) turn-on time

(a) *Delay time* (t_d) : It is the time between the gate current of $0.9 \, I_g$ to reach $0.1 \, I_a$ (anode current). It is also express in time when V_a reaches to $0.9 \, V_a$.

(b) *Rise time* (t_r) : The time taken by anode current to reach from $0.1 \, I_a$ to $0.9 \, I_a$. The rise time is inversely proportional to the magnitude of gate current and its build up rate.

(c) *Spread time* (t_p) : The time taken by anode current reach $0.9 \, I_a$ to I_a is called spread time.

$$t_{on} = t_d + t_p + t_r \approx 1 \text{ to } 4 \text{ } \mu\text{sec}$$

The turn on time can therefore reduced by using higher values of gate currents. The magnitude of gate current is usually 3 to 5 times, the minimum gate current required to turn on SCR.

(2) turn-off time : The SCR can be turn-off by reducing anode current below holding current I_H.

$$t_{off} = t_{rr} + t_{gr}$$

Generally $\boxed{t_{off} > t_{on}}$

Current rating

Average current $= \dfrac{I_{r.m.s.}}{F \cdot F}$

Where $F \cdot F$ is form factor.

Conduction angle	15°	30°	45°	60°	90°	120°	180°
From facter		5.65	3.9812	3.233	2.7781	2.2214	1.878	1.5708

For rectangular wave $F \cdot F$ is less as compared to its value for sine wave for the same conduction angle.

Thyristor protection

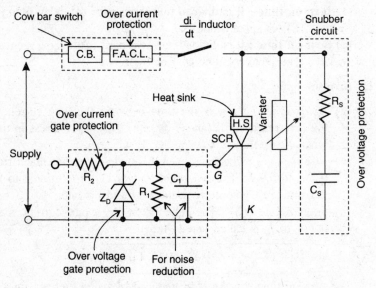

Fig. 2.3

(a) In the above Fig. 2.3 R_S and C_S connected across SCR is used for dv/dt protection. This R_S in series with C_S is called snubber circuit.

(b) di/dt protection can be maintained below acceptable limit by using as small inductor called di/dt inductor. di/dt limit value of SCR are 20-500 A/μ sec.

(c) Over current protection in thyristor circuit is acheived through the use of circuit breaker (C.B.) and fast acting fuses (F.A.C.L.F.). Combination of these are called cow bar switch.

(d) Over voltage across the gate circuit can cause false triggering of the SCR. This can be avoided by using a zener diode across the gate circuit. Over current may rise junction temperature. It can be protected by connecting a resistance R_2 in series with the gate circuit. A capacitor C_1 and a resistor (R_1) are also connected across gate to cathode to by pass the noise signals.

Heating/cooling of thyristor : Thermal resistance is given by,

$$\theta_{12} = \frac{T_1 - T_2}{P_{av}} \ °C/W$$

Fig. 2.3 (a)

$$\theta_{JA} = \theta_{JC} + \theta_{CS} + \theta_{SA}$$

$$P_{av} = \frac{T_1 - T_C}{\theta_{JC}} = \frac{T_C - T_S}{\theta_{CS}} = \frac{T_S - T_A}{\theta_{SA}}$$

$$T_J - T_A = P_{av} (\theta_{JC} + \theta_{CS} + \theta_{SA})$$

Series and parallel operation of Thyristors : SCR connected in series in order to acheive high voltage demand and SCR in parallel in order to acheive high current demand.

$$\text{String efficiency} = \frac{\text{Actual voltage/Current rating of whole SCR}}{\text{Individual voltage/Current rating of one SCR}}$$
$$\times \text{(No of SCR in the string)}$$

For highest string efficiency, SCR must have identical VI characteristic. Those whose leakage reactance is high shares high voltage than other.

De-rating factor

$$\text{DRF} = 1 - \text{String efficiency}$$

With higher value of DRF, more SCR are required and therefore voltage and current shared by each device are lower than their normal rating.

Other member of the thyristor family : (a) PUT (Programmable unjunction transistor) : It is a *pn-pn.* device like an SCR. The basic difference is that gate is connected to *n*-type material near the anode. It is mainly used in time-delay logic and SCR trigger circuits.

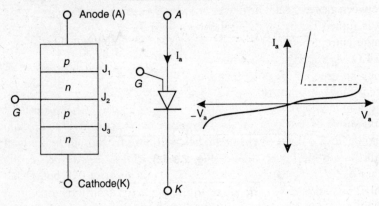

Fig. 2.4

(b) SUS (Silicon unilateral switch) : It is similar to PUT with a diode between gate and cathode. Because of the presence of diode, SUS turns on for a fixed anode to cathode voltage unlike an SCR. It is mainly used in timing, logic and trigger circuit.

(c) SCS (Silicon controlled switch) or (OR Gate) : It has four electrode. It has two gates.

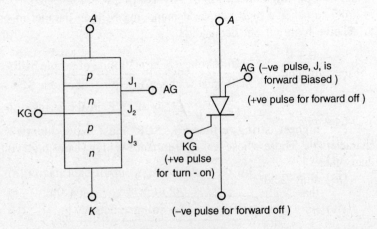

Fig. 2.5 (a) - (b)

It is used for :
(a) timing, logic and triggering circuit
(b) pulse generator
(c) voltage sensor, and
(d) oscillators etc.

(d) Static Induction Thyristors (SITHs) :
SITH is turned on by applying a short +ve pulse
between gate and cathode like an ordinary thyristor.
It is turned off by applying a short –ve pulse of
large current between gate and cathode just like
G.T.O. A SITH has low on-state voltage. These
may be used for medium power converters with
freq. range beyond that useds for G.T.O.

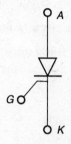

Fig. 2.6

(e) G.T.O. (Gate turn-off) Thyristor : Its switching operation is
similar to (SITHS), except the –ve gate current required to turn-off a
G.T.O. is quite large (20 + 30% of anode current).

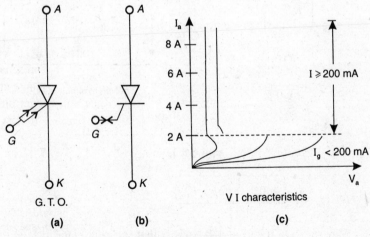

G. T. O.

(a)   (b)   V I characteristics   (c)

Fig. 2.7 (a) - (c)

Advantage of G.T.O. over an SCR

(i) G.T.O. has faster switching speed.
(ii) It has more *di/dt* rating at turn-on.
(iii) It eliminates forced commutation. So it has higher efficiency
than SCR.
(iv) Its surge current capability is comparable with an SCR.

Disadvantage of G.T.O. as compare to SCR :

(i) Magnitude of latching and holding current is more than SCR.
(ii) On state voltage drop is more in a G.T.O.
(iii) Gate drive circuit loss is more than SCR.
(iv) Triggering gates current is higher than SCR.

Firing circuits for thyristors

UJT : It is a three terminal semiconductor device. It consists of a lightly doped N-type silicon bar with a heavily doped P-type material.

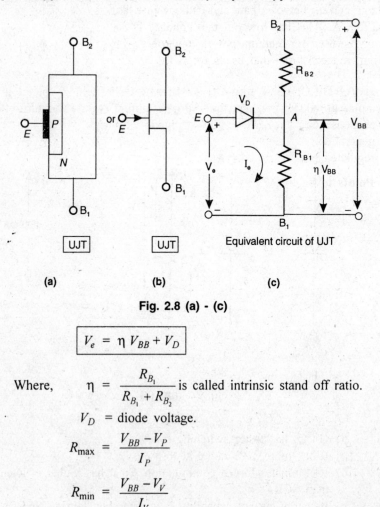

(a)	(b)	(c)

Fig. 2.8 (a) - (c)

$$V_e = \eta V_{BB} + V_D$$

Where, $\eta = \dfrac{R_{B_1}}{R_{B_1} + R_{B_2}}$ is called intrinsic stand off ratio.

V_D = diode voltage.

$$R_{max} = \dfrac{V_{BB} - V_P}{I_P}$$

$$R_{min} = \dfrac{V_{BB} - V_V}{I_V}$$

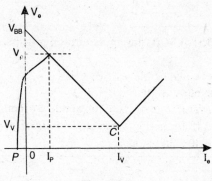

Fig. 2.9

The negative resistance region between peak and valley point (c) gives UJT the switching characteristics for use in SCR triggering circuit.

Application : UJT is used in a variety of circuit application such as phase control, pulse generation, saw tooth generation, size wave generation, switching, timing and trigger circuits, voltage or current regulated supplies, relax oscillator etc.

Points to be noted

(1) An ideal trigger pulse should have short rise time with pulse width greater than turn-on time.

(2) Turn-on time of an SCR in series with RL circuit can be reduced by decreasing inductance L.

(3) In an SCR, anode current flows over a narrow region near the gate during t_d and t_r.

(4) For a pulse transformer, the material used for its core are ferrite with 1:1 turn ratio.

(5) A resistor connected across the gate and cathode of an SCR increases its *dv/dt* rating, holding current, noise immunity.

PROBLEMS

Problem 2.1: An SCR, during its turn-on process, has the following data :

Anode voltage   60 V $V_T = 0V$
Anode current   0 A   100 A

During the turn-on time of 5 μs, the anode current and anode voltage vary linearly. If triggering frequency is 100 Hz. Find the average power loss in the thyristor.

Solution:

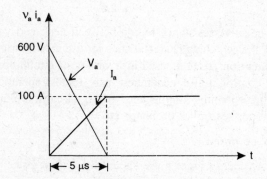

Fig. P. 2.1

Energy loss during turn-on process is given as :

$$= \frac{1}{T} \int_0^{t_{on}} (\text{Instantaneous voltage of anode})$$

$$(\text{Instantaneous current of anode}) \, dt.$$

$$= \frac{1}{T} \int_0^{5\mu s} \left(\frac{I_0}{5 \times 10^{-6}} t \right) \left(600 \, V - \frac{600 \, t}{5 \times 10^{-6}} \right) dt$$

$$= f \int_0^{5\mu s} \left(\frac{100}{5 \times 10^{-6}} t \right) \left(600 \, V - \frac{600 \, t}{5 \times 10^{-6}} \right) dt$$

$$= f \times \frac{100 \times 100}{5 \times 10^{-6}} \int_0^{5\mu s} t \left(6 - \frac{6}{5} \times 10^6 t \right) dt$$

$$= \frac{100 \times 100 \times 100}{5 \times 10^{-6}} \left[6\frac{t^2}{2} - \frac{6}{5} \times 10^6 \frac{t^3}{3} \right]_0^{5\mu s}$$

$$= \frac{10^{12}}{5} \left[6 \times \frac{(25 \times 10^{-12})}{2} - \frac{2}{5} \times 10^6 \times 10^{-18} \times 125 \right]$$

$$= \frac{10^{12}}{5} \left[3 \times 25 \times 10^{-12} - 2 \times 25 \times 10^{-12} \right]$$

$$= \frac{1}{5}[75 - 50] = \frac{25}{5} = 5 \text{ W}$$

So, $\boxed{5\text{ W}}$

Problem 2.2: The gate-cathode characteristic of an SCR is given by $V_g = 0.5 + 8\, I_g$. For a triggering of 400 Hz and duty cycle of 0.1, compute the value of resistance to be connected in series with the gate circuit. The rectangular triggerpulse applied to the gate circuit has an amplitude of 12 V. The thyristor has averages gate power loss of 0.5 watts.

Solution: Gate pulse width is given as :

$$T = \frac{\text{duty cycle}}{\text{triggering freq.}} = \frac{0.1}{400} = 250 \text{ μs}$$

As, $T > 100$ μs, so d.c. data is apply

Hence, $V_g\, I_g = 0.5$ Watts (average power)

or $(0.5 + 8\, I_g)\, I_g = 0.5$

$8\, I_g^2 + 0.5\, I_g - 0.5 = 0$

So, $\boxed{I_g = 0.22 \text{ Amp}}$

During the pulse-on period :

$$\boxed{E_S = R_S\, I_g + V_g}$$

So, $12 = R_S\, I_g + (0.5 + 8\, I_g)$

or $12 = R_S\, I_g + (0.5 + 8 \times 0.22)$

or $12 = R_S \times 0.22 + 2.26$

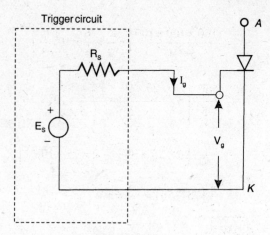

Fig. P. 2.2

or $\qquad R_S = \dfrac{12 - 2.26}{0.22}$

or $\qquad R_S = \dfrac{9.74}{0.22} + 44.27 \,\Omega$

∴ $\qquad \boxed{R_S = 44.27 \,\Omega}$

Problem 2.3: A thyristor data sheet gives 1.5 V and 100 mA as the minimum value of gate-trigger voltage and gate-trigger current respectively. A resistor of 20 Ω is connected across gate-cathode terminals. For a trigger supply voltage of 8 V, compute the value of resistance that should be connected in series with gate circuit in order to ensure turn-on of the device.

Solution:

Here, $\quad E_S = 8V$, $\qquad V_g = 1.5$ V, $\qquad I_1 = \dfrac{1.5\,V}{R_1}$,

$\qquad I_g = 100$ mA, $\qquad R_S = $ Unknown $\qquad R_1 = 20\,\Omega$

Trigger circuit connected to gate-cathode circuit of an SCR.

A resistance R_1 is also connected across gate-cathode terminals, so as to provide an easy path to the flow of leakage current between SCR terminals.

From the given circuit :

$$\frac{E_S - V_g}{R_S} = I_g + I_1$$

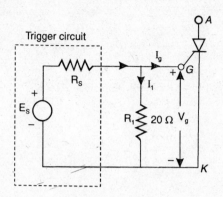

Fig. P. 2.3

or $\dfrac{E_S - V_g}{R_S} = I_g + \dfrac{V_g}{R_1}$

or $\dfrac{8 - 1.5}{R_S} = 100 \text{ mA} + \dfrac{1.5}{20} \text{ A}$

or $\dfrac{6.5}{R_S} = (0.1 + 0.075) \text{ A}$

or $\dfrac{6.5}{R_S} = 0.175$

or $R_S = \dfrac{6.5}{0.175} \Omega$

or $\boxed{R_S = 37.142 \ \Omega}$

Problem 2.4: A thyristor is triggered by a train of pulses of frequency 4 kHz and of duty cycle 0.2 calculate the pulse width. In case average gate power dissipation is 1W. Find the maximum allowable gate power drive.

Solution:

(a) $\quad$ Pulse width $= \dfrac{\text{duty cycle}}{\text{triggering freq.}} \text{ sec} = \dfrac{0.2}{4 \text{ kHz}} \text{ sec}$

$$= \frac{0.2}{4 \times 10^3 \times 10} \sec = \frac{10^{-4}}{2} = 0.5 \times 10^{-4} \sec$$

or | pulse width = 50 μs |

(b) For maximum allowable gate power :

$$\boxed{\frac{P_{gm}T}{T_1} \geq P_{g,ave}}$$

or $\boxed{P_{gm}\delta \geq P_{g,ave}}$

where δ = duty cycle

P_{gm} = max power allowable to the gate

$P_{g,ave}$ = average power

So, $P_{gm} \geq \dfrac{P_{g,ave}}{\delta}$

or $\boxed{P_{gm} = \dfrac{P_{g,ave}}{\delta}}$

or $P_{gm} = \dfrac{1\,W}{0.2}$

So, $\boxed{P_{gm} = 5\ W}$

Problem 2.5 : During turn-off of a thyristor, idealized voltage and current wave form are shown in Fig. P. 2.5. For a triggering frequency of 50 Hz. Find the mean power loss due to turn-off loss. Also obtain the reversed recovery charge.

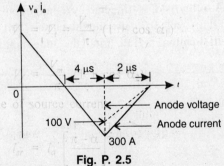

Fig. P. 2.5

Solution: (a) Power loss during turn-off process is given as :

$$= \frac{1}{T}\left[\int_0^{4\mu s} \left(300 - \frac{300t}{4\times 10^{-6}}\right)100\,dt - \int_0^{2\mu s}\left(300t - \frac{300t^2}{2\times 10^{-6}}\right)\times \frac{100\,dt}{2\times 10^{-6}}\right]$$

$$= \frac{100}{T}\left[\left(300t - \frac{300t^2}{4\times 10^{-6}\times 2}\right)\Bigg|_0^{4\mu s} - \left(\frac{300t^2}{10^{-6}\times 2} - \frac{300t^3}{(2\times 10^{-6})^2 \times 3}\right)\Bigg|_0^{2\mu s}\right]$$

$$= \frac{100\times 300}{T}\left[4\times 10^{-6} - \frac{16\times 10^{-12}}{4\times 10^{-6}\times 2} - \frac{(2\times 10^{-6})^2}{2\times 10^{-6}} - \frac{(2\times 10^{-6})^3}{(2\times 10^{-6})^2 \times 3}\right]$$

$$= \frac{100\times 300\times 10^{-6}}{T}\left[(4-2)-\left(2-\frac{2}{3}\right)\right]$$

$$= 50 \times 100 \times 300 \times 10^{-6}\left[2-\frac{4}{3}\right] \qquad \left[\text{where}\frac{1}{T}=f=50\text{ Hz}\right]$$

$$= 1.5\left[\frac{6-4}{3}\right] = 1.5 \times \frac{2}{3} = 1\text{ W}$$

or $\boxed{P = 1\text{ W}}$

(b) Reverse recovery charge is given as :

$$q = \left(\frac{1}{2}\times 300\times 4\ \mu s + \frac{1}{2}\times 300\times 2\mu s\right)C$$
$$= (600\times 10^{-6} + 300\times 10^{-6})\,C = 900\times 10^{-6}\,C$$

$$\boxed{q = 900\ \mu C}$$

Problem 2.6: An SCR has maximum r.m.s. current of 78.5 A. Find its maximum average current rating?

Solution: The r.m.s. current for an SCR is constant what ever the conduction angle may be. But the average current is given by ($I_{r.m.s}/F.F$) where $F.F$ is the form factor of the current wave form. It is maximum at conduction angle $0°$ and minimum at conduction angle $180°$. For a sine wave $F.F$ at $180°$ conduction angle is 1.5708. So max. average current rating of an SCR is :

$$I_{ave\ (max)} = \frac{I_{r.m.s}}{(F.F)_{min}} = \frac{78.5}{1.5708} = 49.97\text{ Amp}$$

So $I_{ave\ (max)}$ = 49.97 Amp

or $\boxed{I_{ave\ (max)} = 50\ \text{Amp}}$

[Note : For the same conduction angle, the F.F of sine wave is higher than, for the rectangular wave.]

Problem 2.7: The specification sheet for an SCR gives maximum r.m.s on-state current as 50 A. If the SCR is used in a resistive circuit, compute its average on-state current rating for conduction angles of 30° and 60° in case current wave form is (1) half sine wave, and (2) rectangular wave.

Solution: (1) For half since wave.

$$I_{ave} = \frac{1}{2\pi} \int_{\theta_1}^{\pi} I_m \sin\theta\ d\theta$$

or $\boxed{I_{ave} = \frac{I_m}{2\pi}(1 + \cos\theta_1)}$

$$I_{r.m.s.} = \left[\frac{1}{2\pi}\int_{\theta_1}^{\pi} I_m^2 \sin^2\theta\ d\theta\right]^{\frac{1}{2}}$$

$$\boxed{I_{r.m.s.} = \left[\frac{I_m}{2\sqrt{\pi}}\left(\pi - \theta_1 + \frac{\sin 2\theta}{2}\right)\right]^{\frac{1}{2}}}$$

(a) For 30° conduction angle, $\theta_1 = 180° - 30° = 150°$

$\therefore$ $I_{ave} = \frac{I_m}{2\pi}\left(1 + (-0.866)\right) = 0.0213\ I_m$

and $I_{r.m.s.} = \frac{I_m}{2\sqrt{\pi}}\left\{\frac{\pi}{6} + \frac{\sin 300°}{2}\right\}^{\frac{1}{2}} = 0.0849\ I_m$

$\therefore$ $F.F = \dfrac{0.0849\ I_m}{0.0213\ I_m} = 3.9818$

$\therefore$ So, for max r.m.s. on-state current of 50 A, average on-state current rating is given as :

$$I_{ave} = \frac{I_{r.m.s.}}{F.F} = \frac{50}{3.98} = 12.55 \text{ Amp}$$

$\therefore$ $\boxed{I_{ave} = 12.55 \text{ Amp}}$

(b) For $60°$ conduction angle, $\theta_1 = 180° - 60° = 120°$

$\therefore$

$$I_{ave} = \frac{I_m}{2\pi} (1 + (-0.5)) = 0.0795 \, I_m$$

$$I_{r.m.s.} = \frac{I_m}{2\sqrt{\pi}} \left[\left(\pi - \frac{2\pi}{3} \right) + \frac{\sin \frac{4\pi}{3}}{2} \right]^{\frac{1}{2}} = 0.21 \, I_m$$

$\therefore$

$$F.F = \frac{I_{r.m.s.}}{I_{ave}} = \frac{0.221 \, I_m}{0.0795 \, I_m} = 2.78$$

So, for 50 A r.m.s. on-state current at $60°$ conduction angle average current is given as :

$$I_{ave} = \frac{50}{2.78} = 17.98 \text{ Amp}$$

$\therefore$ $\boxed{I_{ave} = 17.98 \text{ Amp}}$

(2) For rectangular wave :

$$\text{Conduction angle} = \frac{T}{\eta T} \times 360°$$

or

$$\eta = \frac{360°}{\text{conduction angle}}$$

$$\boxed{I_{ave} = \frac{I \times T}{\eta T} = \frac{T}{\eta}}$$

$$\boxed{I_{r.m.s.} = \left[\frac{I^2 \times T}{\eta T} \right]^{\frac{1}{2}} = \frac{I}{\sqrt{\eta}}}$$

(a) For 30° conduction angle :

$$\eta = \frac{360^o}{30} = 12$$

$$\therefore \qquad \boxed{I_{ave} = \frac{I}{12} \text{ Amp}}$$

$$\boxed{I_{r.m.s.} = \frac{I}{\sqrt{12}} \text{ Amp}}$$

$$\therefore \qquad F.F = \frac{12}{\sqrt{12}} = \sqrt{12} = 3.46$$

So, for 50 Amp, r.m.s. on-state current at 30° conduction angle average on-state current is given as :

$$I_{ave} = \frac{50}{3.46} = 14.43 \text{ Amp}$$

$$\therefore \qquad \boxed{I_{ave} = 14.43 \text{ Amp}}$$

(b For 60° coduction angle :

$$\eta = \frac{360^o}{60^o} = 6$$

$$\therefore \qquad \boxed{I_{ave} = \frac{I}{6} \text{ Amp}}$$

$$\boxed{I_{r.m.s.} = \frac{I}{\sqrt{6}} \text{ Amp}}$$

$$\therefore \qquad F.F = \frac{6}{\sqrt{6}} = \sqrt{6} = 2.45$$

$$\boxed{F.F = 2.45}$$

So, for 60° conduction angle, SCR average on-state current rating at 50 Amp r.m.s. on state-current is given as :

$$I_{ave} = \frac{50}{2.45} = 20.44 \text{ Amp}$$

$$\therefore \qquad \boxed{I_{ave} = 20.44 \text{ Amp}}$$

Problem 2.8: A thyristor is placed between a constant d.c. voltage source of 240 V and resistive load R. The specified limits for di/dt and dv/dt for the SCR are 60 A/µs and 300 V/µs respectively. Determine the values of the di/dt inductor and the snubber circuit parameters. Taking damping ratio as 0.5. Discuss how these parameterms may be modified to suit the working conditions in the circuit.

Solution:

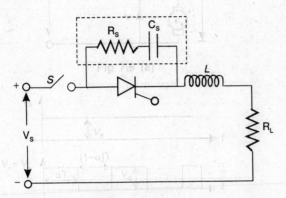

Fig. P. 2.8

When switch S is closed. The capacitor behaves like a short circuit and SCR in the forward blocking state offers a very high resistance. So, after closing the switch, the equivalent circuit is as :

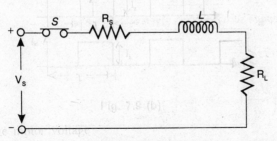

Fig. P. 2.8 (a)

The voltage equation for their circuit is given as :

$$V_S = (R_S + R_L)\, i + L\, \frac{di}{dt}$$

Its solution :

$$i = I\left(1 - e^{-\frac{t}{\tau}}\right) \qquad \qquad ...(1)$$

Where, $\qquad I = \dfrac{V_S}{R_S + R_L}$ and $\tau = \dfrac{L}{R_S + R_L}$

Now, $\qquad \dfrac{di}{dt} = I \cdot e^{-\frac{t}{\tau}} \cdot \dfrac{1}{\tau} = \dfrac{V_S}{R_S + R_L} e^{-\frac{t}{\tau}} \cdot \dfrac{R_S + R_L}{L}$

$$\left(\dfrac{di}{dt}\right) = \dfrac{V_S}{L} e^{-\frac{t}{\tau}}$$

The value of di/dt is maximum when $t = 0$

So, $\qquad \left(\dfrac{di}{dt}\right)_{max} = \dfrac{V_S}{L}$

$\therefore \qquad L = \dfrac{V_S}{\left(\dfrac{di}{dt}\right)_{max}} = \dfrac{240}{\dfrac{60}{10^{-6}}}$

or $\qquad L = \dfrac{240 \times 10^6}{60} = 4\ \mu H$

or $\qquad \boxed{L = 4\ \mu H}$

The voltage across SRC is given by $V_a = R_S . i$

or $\qquad \dfrac{dv_a}{dt} = R_S \dfrac{di}{dt}$

or $\qquad \left(\dfrac{dv_a}{dt}\right)_{max} = R_S \left(\dfrac{di}{dt}\right)_{max}$

$\qquad \left(\dfrac{dv_a}{dt}\right)_{max} = R_S \cdot \dfrac{V_S}{L}$

or $\qquad R_S = \dfrac{L}{V_S}\left(\dfrac{dv_a}{dt}\right)_{max} = \dfrac{4}{240} \times 300$

$$\boxed{R_S = 5\,\Omega}$$

Given, $\xi = 0.5$, the value of capacitor of snubber circuit is given as:

$$R_S = 2\xi\sqrt{\frac{L}{C_S}}$$

or,

$$C_S = \left(\frac{2\xi}{R_S}\right)^2 L = \left(\frac{2\times 0.5}{5}\right)^2 \times 4 \times 10^{-6}$$

$$\boxed{C_S = 0.16\,\mu F}$$

So, computed values of R_S, C_S, L are :

$$\boxed{\begin{array}{l} R_S = 5\,\Omega \\ C_S = 0.16\,\mu t \\ L = 4\,\mu H \end{array}}$$

It is seen that, when switch S is closed. The capacitor C_S is charged and when SCR is turned on, capacitor C_S begains to discharge with maximum current of V_S / R_S and so, the total current through thyristor

is $\left(\dfrac{V_S}{R_S} + \dfrac{V_S}{R_L}\right)$. It should be ensured that this current spike is less than

the peak repetitive current rating (I_{TRM}) of the SCR. Thus if R_S is small, the current spikes will be large. In order to reduce this spikes, R_S is normally taken greater than what is required to limit dv/dt. At the same time, value of C_S is also reduced, so that energy stored in C_S is small. Let, us take $R_S = 8\,\Omega$ and $C_S = 0.12\,\mu F$. The adoption of the new value of R_S, demand a new value of L. So,

$$L = \frac{R_S \cdot V_S}{(dv_a / dt)_{max}}$$

$$= \frac{8 \times 240}{300} = 6.4\,\mu H$$

So, modified values :

$$\boxed{\begin{array}{l} R_S = 8\,\Omega \\ C_S = 0.12\,\mu F \\ L = 6.4\,\mu H \end{array}}$$

Problem 2.9: *R*, *L*, and *C* in an SCR circuit meant for protecting against *dv/dt* and *di/dt* are 4Ω, 6 μH and 6 μF respectively. If the supply voltage to the circuit is 300 V, calculate permissible maximum values of *dv/dt* and *di/dt*.

Solution: When the SCR is on, then equivalent circuit is given as :

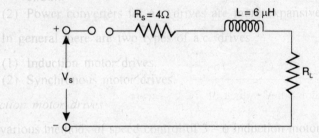

Fig. P. 2.9

SCR on state, equivalent ckt.
From the above circuit:

$$V_S = (R_S + R_L)\, i + L\, \frac{di}{dt}$$

Its soln : $i = I \left(1 - e^{-\frac{t}{\tau}} \right)$

Where, $I = \dfrac{V_S}{R_S + R_L}$ and $\tau = \dfrac{L}{R_S + R_L}$

Now, $\dfrac{di}{dt} = I \cdot e^{-\frac{t}{\tau}} \cdot \dfrac{1}{\tau}$

$$\left(\frac{di}{dt} \right) = \frac{V_S}{L}\, e^{-\frac{t}{\tau}}$$

The value of *di/dt* is maximum when $t = 0$, So,

$$\left(\frac{di}{dt} \right)_{max} = \frac{V_S}{L}$$

or $L = \dfrac{V_S}{\left(\dfrac{di}{dt} \right)_{max}} = 6\ \mu H$

or $\left(\dfrac{di}{dt}\right)_{max} = \dfrac{300 \text{ V}}{6 \, \mu H} = 50 \text{ A}/\mu s$

or $\boxed{\left(\dfrac{di}{dt}\right)_{max} = 50 \text{ A}/\mu s}$

The voltage across SCR is given by :

$$V_a = R_S \cdot i + V_C(0^+)$$

or $\left(\dfrac{dv_a}{dt}\right)_{max} = R_S \cdot \left(\dfrac{dv_a}{dt}\right)_{max} + \left(\dfrac{dv_c}{dt}\right)_{max}(0^+)$

$$\left(\dfrac{dv_a}{dt}\right)_{max} = 4 \times 50 + \dfrac{I_{S\cdot C}}{C} \quad \left[\begin{array}{l} \therefore I_{SC} = C\left(\dfrac{dv_C}{dt}\right)_{max} \\[3mm] \text{or, } I_{SC} = \dfrac{V_S}{R_S} = \dfrac{300}{4} = 75 \text{ A} \end{array}\right]$$

So, $\left(\dfrac{dv_a}{dt}\right)_{max} = \left(200 + \dfrac{75}{6\mu F}\right) V/\mu s = (200 + 12.5) \, V/\mu s$

$$\boxed{\left(\dfrac{dv_a}{dt}\right)_{max} = (212.5) \, V/\mu s}$$

Problem 2.10: Following are the specifications of a thyristor operating from a peak supply of 500 V : Repetitive peak current, $I_P = 250$ A

$$\left(\dfrac{di}{dt}\right)_{max} = 60 \text{ A}/\mu s, \quad \left(\dfrac{dv_a}{dt}\right)_{max} = 200 \text{ V}/\mu s$$

Take a factor of safety of 2 for the three specifications mentioned above. Design a suitable snubber circuit if the minimum load resistance is 20 Ω. Take ξ = 0.65.

Solution: For a factor of safety of 2, the permitted values of I_P, $\left(\dfrac{di}{dt}\right)_{max}$

and $\left(\dfrac{dv_a}{dt}\right)_{max}$ are, $\dfrac{250}{2}, \dfrac{60}{2}$ and $\dfrac{200}{2}$, i.e. 125 A, 30 A/µs and 100 V/µs

respectively. In order to restrict the rate of rise of current beyond specified value (*di/dt*) inductor must be inserted in series with SCR. So

$$L = \frac{V_S}{\left(\dfrac{di}{dt}\right)_{max}} = \frac{500 \times 10^{-6}}{30} \; H$$

So,

$$\boxed{L = 16.66 \; \mu H}$$

and

$$R_S = \frac{L}{V_S}\left(\frac{dv}{dt}\right)_{max} = \frac{16.66 \times 10^{-6}}{500} \times \frac{100}{10^{-6}}$$

$$\boxed{R_S = 3.33 \; \Omega}$$

Before thyristor is turned on, C_S is charged to 500 V. When thyristor is turned on, the peak current through the thyristor is

$$\frac{500}{20} + \frac{500}{3.33} = 175 \; A$$

As, this peak current through SCR is more than the permissible peak current of 125 A, the magnitude of R_S must be increased. Taking R_S as

6 Ω, the peak current through the SCR $= \dfrac{500}{20} + \dfrac{500}{6} = 108.33$ A, less

than allowable peak current. So we choose $R_S = 6 \; \Omega$. Also,

$$C_S = \left(\frac{2\xi}{R_S}\right)^2 L = \left(\frac{2 \times 0.65}{6}\right)^2 \times 16.66 \times 10^{-6}$$

$$\boxed{C_S = 0.78 \; \mu F}$$

The value of C_S must be lowered as to reduce energy stored in the capacitor, so C_S may be taken as 0.5 µF. So, the designed value of snubber circuit is :

$$\boxed{\begin{aligned} R_S &= 6 \; \Omega \\ C_S &= 0.5 \; \mu F \\ L &= 17 \; \mu H \end{aligned}}$$

Problem 2.11: Thyristor shown is Fig. P. 2.11 has I^2t rating of 20 A^2s. It terminal A gets short-circuited to ground, calculate the fault clearance time, so that SCR is not damaged.

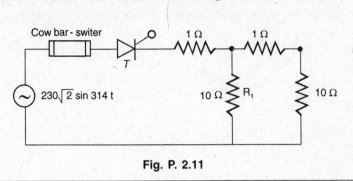

Fig. P. 2.11

Solution: When point A is get short-circuited to the ground, the resistance offered to source $= 1 + \dfrac{10 \times 1}{10 + 1} = \dfrac{21}{11}\,\Omega$. Assuming maximum fault current $= \dfrac{230\sqrt{2} \times 11}{21}\,A$ to remain constant during the short clearance time t_C, we get :

$$\int_0^{t_C} i^2 dt = \int_0^{t_C} \left(\frac{230\sqrt{2} \times 11}{21} \right)^2 dt$$

or $\quad 20\ A^2/s = \left(\dfrac{230\sqrt{2} \times 11}{21} \right)^2 [t]_0^{t_C}$ sec

or $\quad t_C = \dfrac{20}{\left(\dfrac{230\sqrt{2} \times 11}{21} \right)^2}$ sec

or $\quad t_C = \dfrac{20 \times 1000}{(29029.02)}$ ms.

$$\boxed{t_C = 0.6892 \text{ ms}}$$

Problem 2.12: During the turn-off process in a thyristor, the reverse recovery current of 10 A is interrupted in a time interval of 4 μs. The thyristor is connected in series with an inductance of 6 mH with no resistance in the circuit. If the source voltage during turn-off process is –300 V, calculate :

(1) Peak voltage across the thyristor when reverse current is interrupted and

(2) The value of snubber circuit resistance in case snubbber capacitance $C_S = 0.3$ μF and damping ratio is 0.65.

Solution: (1) During the turn-off process, the voltage appears across the thyristor is equal to the sum of voltage across inductor and source voltage.

$$V_p = V_L + V_S = -\left(L\frac{di}{dt}\right) + (-300 \text{ V})$$

$$= -\left(6 \times 10^{-3} \times \frac{10}{4 \times 10^{-6}} \text{ V}\right) + (-300 \text{ V})$$

$$= -(15 \times 10^3 \text{ V}) + (-300 \text{ V}) = -15 \text{ kV} - 0.3 \text{ kV}$$

$$\boxed{V_{rp} = -15.3 \text{ kV across thyristor}}$$

(2) $C_S = 0.3$ μF, $\xi = 0.65$

$$\boxed{R_S = \frac{2\xi}{1}\sqrt{\frac{L}{C_S}}}$$

or $R_S = 2 \times 0.65 \sqrt{\dfrac{6 \times 10^{-3}}{0.3 \times 10^{-6}}} = 1.3 \times \sqrt{2} \times 100$

$$\boxed{R_S = 183.84 \ \Omega}$$

Problem 2.13: A thyristor, having maximum r.m.s. on state current of 45 A, is used in a resistive circuit. Compute its average on-state current rating for half-size wave for conduction angle of $\dfrac{\pi}{3}$ and $\dfrac{\pi}{2}$.

Solution: $I_{ave} = \dfrac{I_m}{2\pi}(1 + \cos\alpha)$ ...(1)

$I_{r.m.s.} = \dfrac{I_m}{2\sqrt{\pi}}\left(\pi - \alpha + \dfrac{1}{2}\sin 2\alpha\right)^{\frac{1}{2}}$ ...(2)

(1)   at $\alpha = \dfrac{\pi}{3}$

$F.F = \dfrac{I_{r.m.s.}}{I_{ave}} = 2.7781$

So, $\boxed{I_{ave} = \dfrac{45}{2.7781} = 16.198 \text{ A}}$

(2)   at $\alpha = \dfrac{\pi}{2}$

$I_{ave} = \dfrac{I_{r.m.s.}}{I_{ave}} = 2.2214$

So, $I_{ave} = \dfrac{45}{2.2214}$

or $\boxed{I_{ave} = 20.25 \text{ A}}$

Problem 2.14: A thyristor is rated to carry full-load current with an allowable case temperature of 100°C, for maximum allowable junction temp of 125°C and thermal resistance between case and ambient as 0.5 °C/W. Find the sink temperature for ambient temperature of 40°C. Take thermal resistance between sink and ambient as 0.4 °C/W.

Solution:

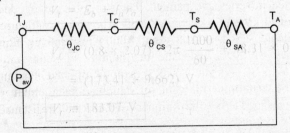

Fig. P. 2.14

Thermal equivalted circuit for an SCR.
From above :

$$P_{ave} = \frac{T_J - T_C}{\theta_{JC}} = \frac{T_C - T_S}{\theta_{CS}} = \frac{T_S - T_A}{\theta_{SA}} = \frac{T_J - T_A}{\theta_{JA}}$$

Where,

$$\boxed{\theta_{JA} = \theta_{JC} + \theta_{CS} + \theta_{SA}}$$

Given,

$$T_J = 125°C, \ T_C = 100°C, \ T_A = 40°C$$
$$\theta_{SA} = 0.4°C/W, \ \theta_{CA} = 0.5°C/W$$
$$\theta_{CA} = \theta_{CS} + \theta_{SA}$$

or

$$\frac{T_C - T_a}{\theta_{CA}} = P_{ave}$$

or $$T_C - 40° = P_{ave} \times \theta_{CA}$$
or $$100° - 40° = P_{ave} \times \theta_{CA}$$

or $$\boxed{60° = P_{ave} \times 0.5}$$ ...(1)

$$T_S - T_a = \theta_{SA} P_{ave}$$

or $$\boxed{T_S - 40° = 0.4 \, P_{ave}}$$ ...(2)

dividing eq. (1) by eq. (2), we get :

$$\frac{60°}{T_S - 40°} = \frac{0.5 \, P_{ave}}{0.4 \, P_{ave}}$$

or $$\frac{60° \times 4}{5} = T_S - 40°$$

or $$48° = T_S - 40°$$

or $$\boxed{T_S = 88 \ °C}$$

Problem 2.15: A Thyristor string is made up of a number of SCR connected in series and parallel. The string has voltage and current rating of 11 kV and 4 kA respectively. The voltage and current rating of available SCR are 1800 V and 1000 A respectively. For a string efficiency of 90 %, calculate the number of series and parallel connected SCR. For these SCR, maximum off-state blocking current is 12 mA. Determine the value of static equilizing circuit resistance for the string. Derive the formula used for their resistance.

Solution: (a) Number of series connected SCR.

$$n_S = \frac{11\,\text{kV}}{1.8\,\text{kA} \times 0.9} = 6.79 \approx 7$$

So, | Series connected SCR = 7 |

Number of parallel connected SCR

$$n_P = \frac{4\,\text{kV}}{1\,\text{kA} \times 0.9} = 4.44 \approx 5$$

So, | Parallel connected SCR = 5 |

(b) Static equilization circuit resistance is used in case of series connected SCR only. Let us consider n thyristor connected in series. Let SCR 1 has minimum leakage current I_{bmn} and each of the remaining $(n - 1)$. SCR have the same leakage current $I_{bmx} > I_{bmn}$.

Note : SCR with lower leakage current blocks more voltage.

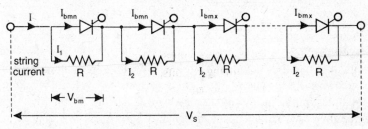

Fig. P. 2.15

$$I_1 = I - I_{bmn} \text{ and } I_2 = I - I_{bmx}$$

Where, I = total string current

Voltage across SCR 1 is $V_{bm} = I_1 R$

Voltage across $(n - 1)$ SCR $= (n - 1)\, I_2 R$

For a string voltage V_S :

$$\begin{aligned}
V_S &= I_1 R + (n - 1)\, I_2 R = V_{bm} + (n - 1)\, R\, (I - I_{bmx}) \\
&= V_{bm} + (n - 1)\, R\, (I_1 - (I_{bmx} - I_{bmn}) \\
&= V_{bm} + (n - 1)\, R I_1 - (n - 1)\, R \cdot \Delta I_b
\end{aligned}$$

Where, | $\Delta I_b = I_{bmx} - I_{bmn}$ |

as $R I_1 = V_{bm},$ | $V_S = n\, V_{bm} - (n - 1)\, R \cdot \Delta I_b$ |

So, $R = \dfrac{n V_{bm} - V_S}{(n - 1)\, \Delta I_b}$

The SCR data sheet contains only maximum blocking current I_{bmx} and rarely ΔI_b. In such a case, it is usual to assume $\Delta I_b = I_{bmx}$ with $I_{bmn} = 0$. Once the value of R is calculated, its power rating is given by

$$P_R = \frac{V_r^2}{R}, \qquad \text{where } V_r = \text{r.m.s. voltage across } R.$$

When $I_{bmx} = 12$ mA, $V_{bm} = 1800$ V, $n_S = 7$ (SCR), $V_S = 11$ kV
then $\Delta I_b = I_{bmx} = 12$ mA

and $R = \dfrac{n V_{bm} - V_S}{(n-1)\, \Delta I_b} = \dfrac{7 \times 1800 - 11000}{(7-1) \times 12 \times 10^{-3}}$

$$= \frac{12600 - 11000}{6 \times 12 \times 10^{-3}} = \frac{1600}{6 \times 12 \times 10^{-3}}$$

$$\boxed{R = 22.22 \text{ k}\Omega}$$

Problem 2.16: Three series connected thyristors, provided with static and dynamic equalizing circuits, have to with stand an off-state voltage of 8 kV. The static equilizing resistance is 8 kV and the dynamic equilizing circuit has $R_C = 40$ Ω and $C = 0.06$ μF. These three thyristors have leakage currents of 25 mA, 23 mA and 22 mA respectively. Determine voltage across each SCR in the off-state and the discharge current of each capacitor at the time of turn on.

Solution: Let I be the string current in the off state.

∴   Voltage across SCR 1 $= (I - 0.025) \times 8000 = V_1$
         Voltage across SCR 2 $= (I - 0.023) \times 8000 = V_2$
         Voltage across SCR 3 $= (I - 0.022) \times 8000 = V_3$

$$V_S = V_1 + V_2 + V_3 + V_4$$

$$8000 = 8000\, (3I - (0.025 + 0.023 + 0.022)$$

or $3I = 0.07 + 1$

or $I = \dfrac{1 + 0.07}{3} = 0.3566$ mA

∴ $\boxed{I = 0.3566 \text{ mA}}$

$\therefore$

$$\boxed{\begin{array}{l} V_1 = 2.653 \text{ kV} \\ V_2 = 2.668 \text{ kV} \\ V_3 = 2.693 \text{ kV} \end{array}}$$

Discharge current through SCR 1 at the time of turn-on :

$$= \frac{V_1}{RC} = \frac{2.653 \text{ kV}}{40 \ \Omega} = 66.325 \text{ Amp}$$

Discharge current through SCR 2 at the time of turn-on :

$$= \frac{V_2}{RC} = \frac{2.668 \text{ kV}}{40 \ \Omega} = 66.70 \text{ Amp}$$

and due to SCR 3 is :

$$= \frac{V_3}{RC} = \frac{2.693 \text{ kV}}{40} = 67.325 \text{ Amp}$$

So, discharge current through SCR 1, SCR 2, SCR 3 are 66.325, 66.70, 67.325 Amp respectively.

Problem 2.17: In a power circuit, four SCR is to be connected in series permissible difference in blocking voltage in 20 V for a maximum difference in their blocking current of 1 mA. Difference in recovery charge is 10 μc. Design suitable equilizing circuit.

Solution:

Blocking voltage is 20 V
Blocking current is 1 mA

So, Resistance in equilizing circuit is equal to :

$$R_S = \frac{20 \text{ V}}{1 \text{ mA}} = 20 \text{ k}\Omega$$

So,

$$\boxed{R_S = 20 \text{ k}\Omega}$$

and capacitance in equilizing circuit is equal to :

$$C_S = \frac{q}{V} = \frac{10 \ \mu C}{20 \text{ V}}$$

or

$$C_S = \frac{10 \times 10^{-6}}{20} = 0.5 \ \mu F$$

So,

$$\boxed{C_S = 0.5 \ \mu F}$$

Problem 2.18: A thyristor in a power converter carries a current of the wave form shown in Fig. P. 2.18. Peak value of the current is 300 A. The static characteristic of the thyristor is given by

$$V_T = 1.05 + 0.95 \times 10^{-3} \, i_T$$

Determine the power loss of the thyristor.

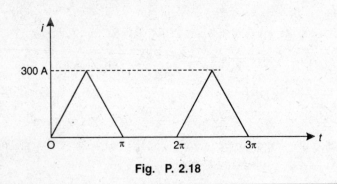

Fig. P. 2.18

Solution: Average value of power loss of thyristor $= \dfrac{1}{2\pi} \displaystyle\int_0^{2\pi} V_T \, i_T \, d\theta$

$$P_{av} = \frac{1}{2\pi} \int_0^{2\pi} (1.05 + 0.95 \times 10^{-3} \, i_T)(i_T) \, d\theta$$

$$= 1.05 \, i_T + 0.95 \times 10^{-3} \, i_T^2$$

Average value of thyristor current $i_{ave} = \dfrac{1}{2} \times \dfrac{300 \times \pi}{2\pi}$

or $i_{ave} = 75$ A

r.m.s. value of thyristor current

$$i_{\text{r.m.s.}} = \left[\frac{1}{2\pi} \int_0^{\pi} i_T^2 \, d\theta \right]^{\frac{1}{2}} = \frac{300}{\sqrt{6}} = 122.47 \text{ Amp}$$

$\therefore$ $P_{ave} = 1.05 \times 75 + 0.95 \times 10^{-3} \times 122.47$

$$= (78.75 + 0.116) \text{ W}$$

$$\boxed{P_{ave} = 78.866 \text{ W}}$$

Problem 2.19: In a power converter, the thyristor current has a wave form of Fig. P. 2.19. Permissible conduction losses are 4 W. The static characteristic of the thyristor can be approximated by

$$V_T = V_{TO} + i_T r_T$$

where $V_{TO} = 1$ V, and $r_T = 26$ mΩ. Determine the permissible value of I_o.

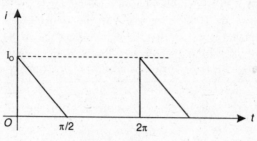

Fig. P. 2.19

Solution: From the wave form of thyristor current

$$i_T = -\frac{2\,I_o}{\pi}t + I_o,\ 0 \le t \le \frac{\pi}{2}$$

Average value of current

$$i_{ave} = \frac{1}{2\pi}\int_0^{\frac{\pi}{2}}\left(-\frac{2\,I_o}{\pi}t + I_o\right)dt = \frac{I_o}{8}$$

r.m.s. value of current

$$i_{r.m.s.} = \left[\frac{1}{2\pi}\int_0^{\frac{\pi}{2}}\left(-\frac{2\,I_o}{\pi}t + I_o\right)^2 dt\right]^{\frac{1}{2}} = \frac{I_o}{\sqrt{12}}$$

The conduction losses

$$P_{ave} = (V_{TO} + r_T i_T)\,i_T = V_{TO}\,i_{T_{ave}} + r_T \cdot i_{T_{r.m.s.}}^2$$

$$4 = 1 \times \frac{I_0}{8} + 26 \times 10^{-3} \times \frac{I_o^2}{\left(\sqrt{12}\right)^2}$$

$$4 = \frac{I_o}{8} + \frac{26}{12} \times 10^{-3} \, I_0^2$$

$\therefore$ $\boxed{I_o = 23 \text{ A}}$ (Peak value)

Problem 2.20: A single phase converter feeds on RL Load having a resistance of 10 Ω in series with an inductance of 30 mH. The converter operates such that the d.c. voltage across the load is 300 V. The thyristor used in the converter has holding current of 300 mA and a delay time of 5 μs. Determine the pulse width of gate current.

Solution: The rate of charge of current at the instant of triggering is

$$= \frac{300}{30 \times 10^{-3}} \text{ A/S} = 10 \times 10^3 \text{ A/S}$$

Time taken for thyristor of current at the instant to reach the holding

current is $= \dfrac{300 \times 10^{-3}}{10 \times 10^{-3}} = 30 \ \mu s$

$\therefore$ Minimum width of gate pulse = 30 μs + 5 μs = 35 μs

---

Problem 2.21: A thyristor has a junction capacitance of 60 PF. The charging current is limited to 20 mA. What is the value of dv/dt? If this is to be limited to 240 V/μs, what capacitance must be connected across the thyristor?

Solution: The junction capacitance = 60 PF.
and charging current   = 20 mA.

The value of $\dfrac{dv}{dt}$ is given as :

$$\frac{dv}{dt} = \frac{20 \times 10^{-3}}{60 \times 10^{-12}} = 0.333 \times 10^3 \text{ V/}\mu s$$

$$= 333.3 \text{ V/}\mu s$$

If this is to be limited by 240 V/μs, then
the total capacitance is = 60 PF

Problem 2.22: Consider the thyristor circuit of Fig. P. 2.22. The thyristor is given a triggering pulse after every 10 ms. Calculate the duration for which the thyristor remains on after each triggering pulse. Assume ideal devices and explain briefly the basis.

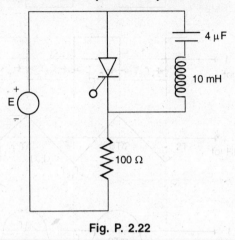

Fig. P. 2.22

Solution: Since the circuit is a series resonant circut. So the oscillating frequency is given as :

$$f = \frac{1}{2\pi\sqrt{LC}} = \frac{1}{T}$$

or

$$\frac{T}{2} = \pi\sqrt{LC} = \pi\sqrt{4\times10^{-6}\times10\times10^{-3}} = 0.63 \text{ ms}$$

off time $CR = 4\times10^{-6}\times100 = 0.4$ ms

∴ Total on-off time is given as = 0.63 ms + 0.4 ms = 1.03 ms

So, the thyristor remains 'on' for the duration of 10 × 0.63 ms between the triggering pulses. When the thyristor is off, the capacitor is charged to a value of 'E'. When the thyristor is triggered oscillation are started in the reasonant circuit which switches "on and off" the thyristor during positive and negative cycle of oscillation.

The capacitor is fully discharged during the off period, i.e., in 0.4 ms and charged within the period of 0.63 ms.

Problem 2.23: A thyristor having maximum r.m.s. on state current of 60 A, is used in a resistive circuit. Compute its average on-state current rating of half sine wave for conduction angle of $\dfrac{\pi}{3}$ and $\dfrac{\pi}{2}$.

Solution: Average value of current is given as

$$I_{ave} = \frac{I_m}{2\pi}(1 + \cos \alpha) \qquad \qquad ...(1)$$

and r.m.s. value of current is given as

$$I_{r.m.s.} = \frac{I_m}{2\sqrt{\pi}}\left((\pi - \alpha) + \frac{1}{2}\sin 2\alpha\right)^{\frac{1}{2}} \qquad ...(2)$$

$\therefore$ (1) At $\alpha = \dfrac{\pi}{3}$

Form factor $(\overset{*}{F}F) = \dfrac{I_{r.m.s.}}{I_{ave}} = 2.7781$

So, $\boxed{I_{ave} = \dfrac{60}{2.7781} = 21.59 \text{ Amp}}$

(2)   At $\alpha = \dfrac{\pi}{2}$

$$F.F = \frac{I_{r.m.s.}}{I_{ave}} = 2.2214$$

$$I_{ave} = \frac{60}{2.2214}$$

$$\boxed{I_{ave} = 27 \text{ Amp}}$$

CHAPTER 3

COMMUTATION

Commutation : Commutation of thyristor is defined as the process of turning-off a thyristor. The thyristor turn-off requires (i) its anode current should falls below the holding current, (ii) a reverse voltage is applied to thyristor for a sufficient time to enable it to recover to blocking state.

There are various type of commutation techniques. They are

(a) Class A commutation (Load commutation)

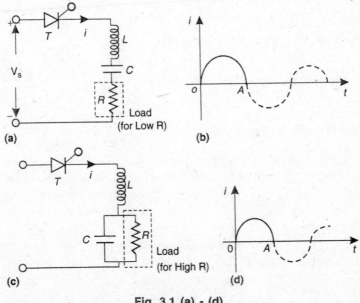

(a)

(b)

(c)

(d)

Fig. 3.1 (a) - (d)

Class A or load commutation (a) series capacitor, (b) shunt capacitor.

Class A or Load commutation is also known as self-commutation or resonant commutation. The essential requirement for both the circuits is that the overall circuit must be under damped.

$$t_o = \text{Conduction time of the thyristor} = \pi\sqrt{LC}$$

(b) Class B commutation (Resonant-pulse commutation)

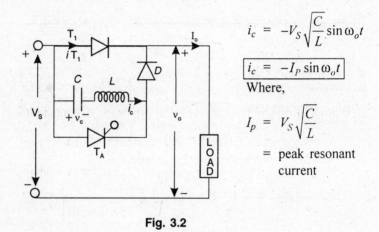

$$i_c = -V_S\sqrt{\frac{C}{L}}\sin\omega_o t$$

$$i_c = -I_P\sin\omega_o t$$

Where,

$$I_p = V_S\sqrt{\frac{C}{L}}$$

= peak resonant current

Fig. 3.2

Maximum current through thyristor $T_1 = \dfrac{V_S}{R}$

Maximum current through thyristor $T_A = V_S\dfrac{\sqrt{C}}{L}$

In this commutation, peak resonant current I_P must be greater than load current I_o for reliable commutation. In this commutation, thyristor is commutated by the gradual build up of resonant current in the reversed direction. So it is also called resonant pulse commutation or current commutation.

Both class A and B commutation circuits are called self commutating circuits as SCR goes off automatically.

Class A commutation, has a limited control range and load variations affect the operation. This problem is eliminated in class B commutation.

(c) Class C commutation (Complementary Commutation)

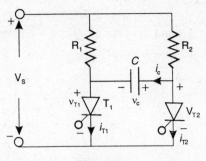

Fig. 3.3

In this type of commutation a thyristor carrying load current is commutated by transferring its load current to another incoming thyristor. In this Figure, firing of SCR T_1 commutates T_2 and subsequently, firing of SCR T_2 would turn off T_1.

$$i_C(t) = \frac{V_S}{R_2} \cdot e^{-\frac{t}{R_2 C}}$$

and

$$V_C(t) = V_S\left(1 - e^{-\frac{t}{R_2 C}}\right)$$

$$t_{C1} = R_1 C \ln(2)$$

or

$$C = \frac{t_{C1}}{R_1 \ln 2}$$

and

$$t_{C2} = R_1 C \ln(2)$$

or

$$C = \frac{t_{C2}}{R_2 \ln 2}$$

In practice the value of C is larger than this. However, if the load contains an inductance like a motor, the value of C can be reduced to the calculated value.

(d) Class D Commutation (Impulse commutation)

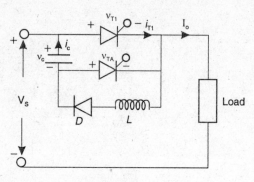

Fig. 3.4

T_1 = main thyristor.

T_A = Auxiliary thyristor.

Initially, main thyristor T_1 and auxiliary thyristor T_A are off and capacitor is assumed charged to voltage V_S with upper plate positive. When T_1 is turned on at $t = 0$, source voltage V_S is applied across load and load current I_o begins to flow which is assumed to remain constant.

$$i_C = V_S \sqrt{\frac{C}{L}} \ \sin \ \omega_0 t = I_P \sin \omega_0 t$$

When $\omega_0 t = \pi$, $i_C = 0$, $i_{T1} = I_0 + I_P \sin \omega_0 \ t \ \left(0 < t < \dfrac{A}{\omega_0} \right)$ capacitor

voltage changes from $+ V_S$ to $- V_S$.

At $\omega_0 t = \pi$, $i_C = 0$, $i_{T1} = I_o$ and $V_C = - V_S$.

At time t_1, T_A is turned on. Immediately after T_A is on capacitor voltage V_S applies a reverse voltage across main thyristor T_1. So that V_{T1} $= - V_S$ at t_1 and SCR T_1 is turned off and $i_{T1} = 0$. Circuit turn-off time for main thyristor.

$$t_C = C \ \frac{V_S}{I_o} \quad \text{where} \quad I_0 = C \ \frac{V_S}{t_C}$$

The circuit turn-off time for auxiliary thyristor

$$t_{C1} = \frac{\pi}{2\omega_0} \quad \text{where} \quad \omega_0 = \frac{1}{\sqrt{LC}}$$

This type of commutation is also known as voltage commutation or auxilary commutation.

(e)   Class E commutation (External pulse commutation)

In this type of commutation, a pulse of current is obtained from a separate voltage source to turn off the conducting SCR. The peak value of this current must be more than the load current.

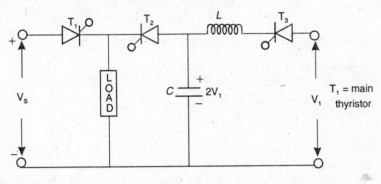

Fig. 3.5

External pulse commutation circuit.

When thyristor T_3 is turned on at time $t = 0$; L, C, T_3 and V_1 form an oscillatory circuit. Therefore C is charged to a voltage $+2 V_1$ with upper plate positive at $t = \pi\sqrt{LC}$. As oscillatory current falls to zero thyristor T_3 get commutated. For turning off the main thyristor T_1, thyristor T_2 is turned on. With T_2 on, T_1 is subjected to a reverse voltage equal to $(V_S - 2V_1)$ and T_1 is therefore turned off.

(f) Class F commutation (Line commutation)

This type of commutation is also known as natural commutation. This method of commutation is applied to phase-controlled converters, line commutated inverters, a.c. voltage controllers and step-down cycloconverter.

Here the thyristor carrying the load current is reverse

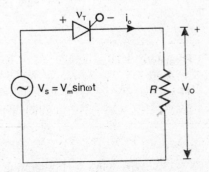

Fig. 3.6 Line commutation circuit

biased by the a.c. source voltage and the device is turned off when anode current falls below the holding current.

The SCR turns off automatically when the size wave goes through the negative cycle. The only constraint is that the duration of the negative half cycle must be greater than SCR turn-off time, i.e.

$$t_C = \frac{\pi}{\omega} > t_q \ \text{(thyristor turn-off time)}$$

Another method of classification of thyristor are

(1) Load commutation : Class A.
(2) Force commutation : Class B, Class C and Class D.
(3) External commutation : Class E.
(4) Natural commutation : Class F.

Note :

(1) In load commutation, L and C are connected in series with the load or C in parallel with the load such that over all load circuit is under damped. Load commutation is commonly employed in series inverter.

(2) In force commutation, the commutating component L and C do not carry load current continuously. So class B, C and D are classified as forced commutation techniques. Force commutation is commonly employed in d.c. chopper and inverter.

(3) D and L across main thyristor accelerate the discharging of the capacitor see Class B and Class D commutation.

PROBLEMS

Problem 3.1: A number of SCR, each with a rating of 2000 V and 50 A are to be used in series-parallel combination in a circuit to handle 11 kV and 400 A. For a derating factor of 0.15, calculate the number of SCR in series and parallel limits.

Solution:

String sefficiency $= 1 -$ derating factor $= 1 - 0.15 = 0.85$

No. of SCR connected in series is given as :

$$N_S = \frac{11000}{2000 \times 0.85} = 6.47 \approx 7$$

So, $\boxed{N_S = 7 \text{ SCR}}$ in series

No. of SCR connected in parallel is given as :

$$N_P = \frac{400}{50 \times 0.85} = 9.4 \approx 10$$

So, $\boxed{N_P = 10 \text{ SCR}}$ in parallel

Problem 3.2: A circuit employing parallel-resonance turn-off (or class-B commutation) circuit has $C = 50\ \mu s$, $L = 20\ \mu H$, $V_S = 200$ V, and initial voltage across capacitor is 200 V. Determine the circuit turn-off time for main thyristor for load $R = 1.5\ \Omega$.

Solution:

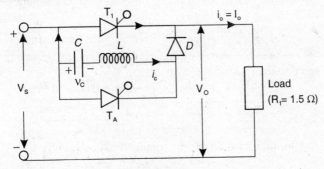

Fig. P. 3.2

Main thyristor, T_1 is turned off when

$$V_S \sqrt{\frac{C}{L}}\ \sin \omega_0 (t_3 - t_2) = I_o$$

or $\qquad \omega_0 (t_3 - t_2) = \sin^{-1} \left(\frac{I_o}{I_P} \right)$

where $I_P = V_S \sqrt{\dfrac{C}{L}}$ = Peak resonant current

Circuit turn-off time for main thyristor

$$t_c = t_4 - t_3 = C \frac{V_{ab}}{I_O}$$

where V_{ab} is the magnitude of reverse voltage across main thyristor T_1, when it gets commutated and is given by

$$V_{ab} = V_S \cos \omega_0 (t_3 - t_2)$$

So, $I_P = V_S \sqrt{\dfrac{C}{L}} = 200 \sqrt{\dfrac{50 \times 10^{-6}}{20 \times 10^{-6}}}$

$$\boxed{I_P = 316.22 \text{ Amp}}$$

$$\boxed{I_o = \frac{V_O}{R_L} = \frac{200}{1.5} = 133.33 \text{ Amp}}$$

So, $\omega_0 (t_3 - t_2) = \sin^{-1} \left(\dfrac{I_O}{I_P} \right) = \sin^{-1} \left(\dfrac{133.33}{316.22} \right) = 24.93°$

$\therefore$ $\boxed{\omega_0 (t_3 - t_2) = 24.93°}$

or $\cos \omega_0 (t_3 - t_2) = 0.906$

$\therefore$ $V_{ab} = V_S \cos \omega_0 (t_3 - t_2) = 200 \times 0.906$

$$\boxed{V_{ab} = 181.35 \text{ V}}$$

$\therefore$ Circuit turn-off time for main thyristor is

$$t_C = t_4 - t_3 = C \frac{V_{ab}}{I_O} \text{ sec} = 50 \times 10^{-6} \times \frac{181.35}{133.33} \text{ sec} = 68 \times 10^{-6} \text{ sec}$$

$$\boxed{t_C = 68 \text{ μsec}}$$

Problem 3.3: For the circuit in Fig. P. 3.3 supply voltage $V_S = 230$ V d.c., load current $I_O = 200$ A, circuit turn-off time for main thyristor = 25 μs and reversal current is limited to 150 % of I_o. Determine the value of commutating components C and L.

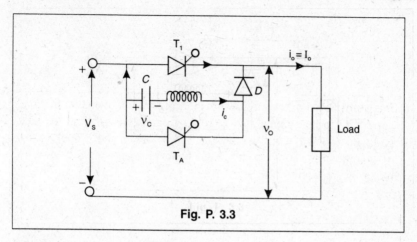

Fig. P. 3.3

Solution: Given, $I_o = 200$ A

and $\qquad I_P = 150\%$ of $I_o = 1.5 \times 200 = 300$ A

So, $\qquad \boxed{I_P = 300 \text{ Amp}}$

So, $\omega_0 (t_3 - t_2) = \sin^{-1}\left(\dfrac{I_o}{I_P}\right) = \sin^{-1}\left(\dfrac{1}{1.5}\right)$

$\qquad \boxed{\omega_0 (t_3 - t_2) = 41.81^\circ}$

$\qquad V_{ab} = V_S \cos (\omega_0 (t_3 - t_2)) = 230 \times \cos 41.81^\circ$

$\qquad \boxed{V_{ab} = 171.43 \text{ V}}$

$\therefore \qquad t_C = C \cdot \dfrac{V_{ab}}{I_o}$

or $\qquad C = \dfrac{t_C \cdot I_o}{V_{ab}} = \dfrac{25 \times 10^{-6} \times 200}{171.43}$

$\because \qquad \boxed{C = 29.16 \ \mu\text{F}}$

$\qquad I_P = V_S \sqrt{\dfrac{C}{L}}$

$\therefore \qquad L = \dfrac{V_S^2 C}{I_P^2} = \dfrac{(230)^2}{(300)^2} \times 29.16 \times 10^{-6} \text{ H}$

or $L = 17.13 \times 10^{-6}$ H

or $\boxed{L = 17.13 \ \mu H}$

Problem 3.4: For the circuit shown in Fig. P. 3.4 given that the load current I_O to be commutated is 10 A. Circuit turn-off time required is 40 μs and the supply voltage is 100 V, obtain the proper values of commutating components. Take peak resonant current equal to twice the load current.

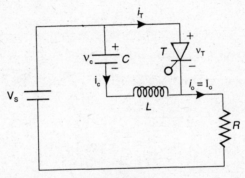

Fig. P. 3.4

Solution: $I_o = 10$ A

$I_P = 2 I_o = 20$ A

$\therefore \quad \omega_0 (t_3 - t_2) = \sin^{-1}\left(\dfrac{I_o}{I_P}\right) = \sin^{-1}\left(\dfrac{10}{20}\right) = \sin^{-1}(0.5) = 30^\circ$

$\therefore \quad \omega_0 (t_3 - t_2) = 30^\circ$

$\therefore \qquad\qquad V_{ab} = V_S \cos \omega_o (t_3 - t_2)$

$\boxed{V_{ab} = 100 \cos 30^\circ}$

Given, $I_P = 2.5 I_o$

$\therefore \quad \omega_o (t_3 - t_2) = \sin^{-1}\left(\dfrac{I_o}{I_P}\right) = \sin^{-1}\left(\dfrac{1}{2.5}\right)$

$\boxed{\omega_o (t_3 - t_2) = 23.57^\circ}$

$\therefore \qquad\qquad V_{ab} = V_S \cos \omega_o (t_3 - t_2) = V_S \cos 23.57^\circ$

$$V_{ab} = 0.916\ V_S$$

$$I_P = V_S \sqrt{\frac{C}{L}}$$

or $\quad 2.5\ I_o = V_S \sqrt{\frac{C}{L}}$

$$I_o = \frac{V_S}{2.5} \sqrt{\frac{4 \times 10^{-6}}{18 \times 10^{-6}}} = \frac{V_S}{2.5} \sqrt{\frac{4 \times 10^{-6}}{18 \times 10^{-6}}}$$

$$\boxed{I_o = \frac{V_S}{2.5} 0.47}$$

∴ $\quad t_C = C\ \dfrac{V_{ab}}{I_o}$

$$t_C = 4 \times 10^{-6} \times \frac{0.916\ V_S \times 2.5}{0.47\ V_S} = 19.48 \times 10^{-6}$$

or $\quad \boxed{t_C = 19.48\ \mu F}$

Conduction time for auxiliary thyristor

$$= \frac{\pi}{\omega_o} = \frac{\pi}{\dfrac{1}{\sqrt{LC}}} = \pi \sqrt{LC} = \pi \sqrt{18 \times 4 \times 10^{-6} \times 10^{-6}}$$

$$= \pi \times 8.48 \times 10^{-6} = 26.65\ \mu\ \text{sec}$$

or $\quad V_{ab} = 100 \times 0.866$

$$\boxed{V_{ab} = 86.6\ V}$$

∵ $\quad t_C = C \cdot \dfrac{V_{ab}}{I_0}$

or $\quad 40 \times 10^{-6} = C \times \dfrac{86.6}{10}$

or $\quad C = \dfrac{40 \times 10^{-6}}{8.66}$

or $C = 4.618 \times 10^{-6}$

or $\boxed{C = 4.618 \ \mu F}$

$\because$ $I_P = V_S \sqrt{\dfrac{C}{L}}$

or $20 = 100 \sqrt{\dfrac{4.618 \ \mu F}{L}}$

or $L = 4.618 \times 25 \ \mu H$

or $\boxed{L = 115.47 \ \mu H}$

Problem 3.5: For the circuit shown in Fig. P. 3.5 thyristor current $= 2.5$ times the constant load current, $L = 18 \ \mu H$ and $C = 4 \ \mu F$. Find the time elapsed from the instant thyristor is turned on to the instant it gets turned off.

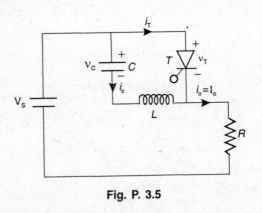

Fig. P. 3.5

Solution: Time elapsed from the instant thyristor is turned on to the instant it gets turned off is equal to :

$$t' = t_C + \frac{\pi}{2\sqrt{\dfrac{1}{LC}}} = t_C + \frac{\pi\sqrt{LC}}{2}$$

$$t' = \left(19.48 + \frac{26.65}{2}\right) \mu \ sec$$

$$t' = 32.805 \ \mu \ sec$$

Problem 3.6: For current commutated circuit of Fig. P. 3.6 : $V_S =$ 230 V, $L = 16 \ \mu H$, and $C = 5 \ \mu F$. Capacitor is initially charged to voltage V_S with left hand plate positive. Auxiliary thyristor T_A is turned on at $t = 0$. Find the total time for which capacitor current i_C exists. The peak resonant current is 1.5 times the full-load current.

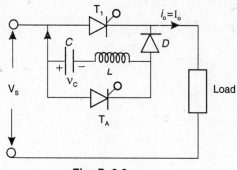

Fig. P. 3.6

Solution:

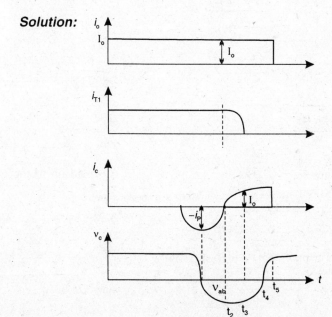

Fig. P. 3.6 (a)

From the Fig. P. 3.6 (a)

$$t_5 - t_3 = C\left(\frac{V_S + V_{ab}}{I_o}\right)$$

$$\omega_o \, (t_3 - t_2) = \sin^{-1}\left(\frac{I_o}{I_P}\right) = \sin^{-1}\left(\frac{1}{1.5}\right)$$

$$\omega_o \, (t_3 - t_2) = 41.81°$$

$$V_{ab} = V_S \cos \omega_o \, (t_3 - t_2) = 230 \cos 41.81°$$

$$V_{ab} = 171.43 \text{ V}$$

$$I_o = \frac{I_P}{1.5} = \frac{V_S \sqrt{\dfrac{C}{L}}}{1.5} = \frac{230 \sqrt{\dfrac{5}{16}}}{1.5}$$

$$I_o = \frac{128.57}{1.5} \text{ Amp}$$

or

$$I_o = 85.71 \text{ Amp}$$

$$\therefore \quad (t_5 - t_3) = C \frac{V_S + V_{ab}}{I_o} = 5 \times 10^{-6} \times \frac{(230 + 171.43)}{85.71}$$

$$(t_5 - t_3) = 23.41 \, \mu \text{ sec}$$

or $\quad \omega_0 \, (t_3 - t_2) = \dfrac{41.81 \times \pi}{180}$

or $\quad (t_3 - t_2) = 0.72 \sqrt{LC} = 0.72 \times \sqrt{16 \times 5} \, \mu \text{ sec}$

$$(t_3 - t_2) = 6.52 \, \mu \text{ sec}$$

Conduction time for auxiliary thyristor

$$= \frac{\pi}{\omega_o} \sec = \pi \sqrt{LC} = \pi \sqrt{16 \times 5} \, \mu \text{ sec} = 28.09 \, \mu \text{ sec}$$

So, total time for which capacitor current i_C exists is equal to :

$$= (28.09 + 6.52 + 23.41) \text{ }\mu \text{ sec}$$

$$\boxed{t_T = 58.029 \text{ }\mu \text{ sec}}$$

Problem 3.7: In the circuit of Fig. P. 3.7 employing complementary commutation ; $V_S = 200$ V, $R_1 = 20$ Ω and $R_2 = 100$ Ω. Determine the minimum value of C so that thyristors do not get turned on due to re-applied dv/dt. Each SCR has a minimum charging current of 4 mA to turn it on and its junction capacitance is 20 PF.

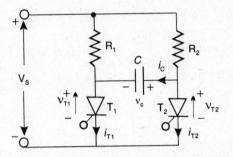

Fig. P. 3.7: Class C complementary commutation

Solution: In class C commutation, when T_1 is to be turned off, T_2 is triggered. If T_2 is turned on at t_1, then capacitor voltage V_C applies a reverse potential V_S across SCR T_1 and turn it off. From the above circuit:

Applying KVL: $R_1 \, i_C + \dfrac{1}{C}\displaystyle\int i_C \, dt = V_S$

or $R_1 I_C(s) + \dfrac{1}{C}\left[\dfrac{I_C(s)}{S} - \dfrac{C V_S}{S}\right] = \dfrac{V_S}{S}$ [In laplace form]

Its solution gives :

$$i_C(t) = \frac{2V_S}{R_1} \, e^{-\frac{t}{R_1 C}}$$

Voltage across capacitor is

$$V_C(t) = \left[\frac{1}{C}\int_0^t i_C \, dt + V_S\right]$$

$$V_C(t) = V_S \left[2e^{-\frac{t}{R_1 C}} - 1 \right]$$

$$\left[\frac{dv_C}{dt} \right] = \frac{2V_S\, e^{-\frac{t}{R_1 C}}}{R_1 C}$$

or $\quad \boxed{\left[\dfrac{dv_C}{dt} \right]_{t=0} = \dfrac{2V_S}{R_1 C}} \qquad\qquad \ldots(1)$

The rate of rise of capacitor voltage V_C across may be large. In case

SCR charging current $C_J \cdot \left(\dfrac{dv_C}{dt} \right)_{t=0}$ happens to be equal to 4 mA, SCR

will get turned on. Here C_J is the junction capacitor of SCR.

So, $\qquad\qquad C_J \left(\dfrac{dv_C}{dt} \right)_{t=0} = 4$ mA

Substituting the value of $\left(\dfrac{dv_C}{dt} \right)_{t=0}$ in eq. (1)

we get :

$$C_J \frac{2V_S}{R_1 C} = 4 \text{ mA}$$

or $\quad \dfrac{20 \times 10^{-12} \times 2 \times 200}{20 \times C} = 4 \times 10^{-3}$

or $\qquad\qquad C = \dfrac{400}{4} \times 10^{-9}$ F

or $\qquad\qquad \boxed{C = 0.1\ \mu F}$

Problem 3.8: The circuit shown in Fig. P. 3.8 is initially relaxed. The thyristor T is turned on at $t = 0$. Determine:
(a) Conduction time of thyristor.
(b) Voltage across thyristor and capacitor after SCR is turned off calculate these values for $L = 10$ mH and $C = 20\ \mu F$ and $V_S = 220$ V.

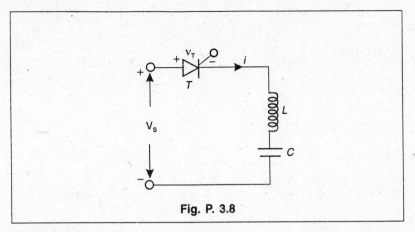

Fig. P. 3.8

Solution: When thyristor is turned on, it behaves like a diode. So applying KVL for this circuit gives

$$L \frac{di}{dt} + \frac{1}{C} \int i \, dt = V_S$$

Its solution is given by

$$i(t) = V_S \sqrt{\frac{C}{L}} \sin \omega_o t$$

Here $\omega_o = \dfrac{1}{\sqrt{LC}}$ = resonant frequency

Voltage across capacitor is :
$$V_C(t) = V_S(1 - \cos \omega_o t) \qquad \qquad ...(1)$$

At time $t = t_o = \dfrac{\pi}{\omega_o}$, $i(t) = 0$ and $V_C(t) = -2V_S$

Here t_o = conduction time of the thyristor = $\pi\sqrt{LC}$

i.e. $\boxed{t_0 = \pi\sqrt{LC}} \qquad \qquad ...(2)$

Voltage V_T across thyristor during conduction time to is zero. When it stops conducting, $V_T = -2V_S + V_S = -V_S$. This shows that SCR is subjected to a reverse voltage of V_S which helps in its recovery.

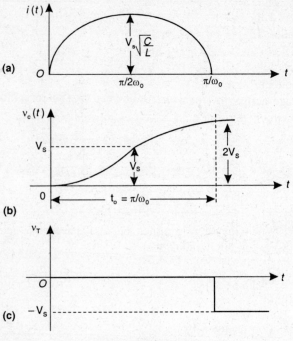

Fig. P. 3.8 (a) - (c)

Resonant frequency $\omega_o = \dfrac{1}{\sqrt{LC}}$

$\therefore \qquad \omega_o = \dfrac{1}{\sqrt{10 \times 10^{-3} \times 20 \times 10^{-6}}} = 0.223606 \times 10^4$ rad/s

$$\boxed{\omega_o = 2236.06 \text{ rad/sec}}$$

Conduction time of thyristor

$$t_o = \frac{\pi}{\omega_o} = \frac{\pi \sqrt{20}}{10^4} = 1.4 \times 10^{-3}$$

$$\boxed{t_o = 1.4 \text{ ms}}$$

Voltage across thyristor after it is turned off is equal to the voltage across SCR, i.e.

$$= -V_S = -220 \text{ V}$$

Problem 3.9: Circuit of Fig. P. 3.9 employing resonant-pulse commutation (or class-B commutation) has $C = 30$ µF and $L = 10$ µH. Initial voltage across capacitor is $V_S = 230$ V. For a constant load current of 300 A, calculate:

(a) Conduction time for the auxiliary thyristor.
(b) Voltage across the main thyristor when it gets commutated.
(c) The circuit turn-off time for the main thyristor.

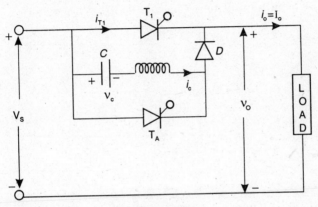

Fig. P. 3.9

Solution: Peak value of resonant current

$$I_P = V_S \sqrt{\frac{C}{L}} = 230 \sqrt{\frac{23 \times 10^{-6}}{10 \times 10^{-6}}}$$

$$\boxed{I_P = 398.37 \text{ A}}$$

Resonant frequency

$$\omega_o = \frac{1}{\sqrt{LC}} = \frac{10^6}{\sqrt{300}} = 57.73 \times 10^3 \text{ rad/s}$$

or, $\boxed{\omega_o = 57.73 \times 10^3 \text{ rad/s}}$

(a) Conduction time for auxiliary thyristance

$$= \frac{\pi}{\omega_o} = \frac{\pi}{57.73 \times 10^3} = 54.41 \times 10^{-6} \text{ s} = 54.41 \text{ µs}$$

(b) Since

$$\omega_o \, (t_3 - t_2) = \sin^{-1} \left(\frac{300}{398.37} \right) = 48.85° = 0.852 \text{ rad}$$

Voltage across main thyristor, when it gets turned off is given by
$$V_{ab} = V_S \cos \omega_o \, (t_3 - t_2) = 230 \cos (48.85°)$$

$$\boxed{V_{ab} = 151.34 \text{ V}}$$

(c) Circuit turn-off time for main thyristor is given as

$$t_C = t_4 - t_3 = C \, \frac{V_{ab}}{I_o}$$

or
$$t_C = 30 \times 10^{-6} \times \frac{151.34}{300}$$

$$t_C = 15.13 \times 10^{-6} \text{ sec}$$

or
$$\boxed{t_C = 15.13 \; \mu \text{ sec}}$$

Problem 3.10: Circuit of Fig. P. 3.10, employing class C commutation, has $V_S = 200$ V, $R_1 = 10 \, \Omega$, $R_2 = 100 \, \Omega$, Determine:
(a) Peak value of current through thyristor T_1 and T_2.
(b) Value of capacitor C if each thyristor has turn-off time of 30 μs. Take a factor of safety 2.

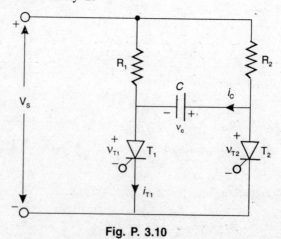

Fig. P. 3.10

Solution: (a) Peak value of current through thristor T_1

$$= V_S \left[\frac{1}{R_1} + \frac{2}{R_2} \right] = 200 \left[\frac{1}{10} + \frac{2}{100} \right] = 24 \text{ A}$$

and peak value of current through thyristor T_2 is

$$= V_S \left[\frac{2}{R_1} + \frac{1}{R_2} \right] = 200 \left[\frac{2}{10} + \frac{1}{100} \right] = 42 \text{ A}$$

(b) Thyristor turn-off time is given by

$$t_{cn} = R_n C \ln (2)$$

or $$C = \frac{t_{cn}}{R_n \ln (2)}$$

Since factor of safety is given 2.

So, $$C' = 2C = \frac{2 \times t_{C_1}}{R_1 \ln (2)} = \frac{2 \times 30 \times 10^{-6}}{100 \ln (2)} = 8.656 \text{ }\mu\text{F}.$$

For resistance of $R_2 = 100 \text{ }\Omega$

$$C' = \frac{2 \times t_{C_2}}{R_2 \ln (2)} = \frac{2 \times 30 \times 10^{-6}}{100 \times \ln (2)} = 0.865 \text{ }\mu\text{F}$$

So, we choose a capacitor of large size of 8.656 μF.

Problem 3.11: In the circuit shown in Fig. P. 3.11. SCR is forced commutated by circuitry not shown in Figure. Compute the minimum value of C, so that SCR does not get turned on due to re-applied dv/dt. The SCR has minimum charging current of 5 mA to turn it on and its junction capacitance is 25 pF.

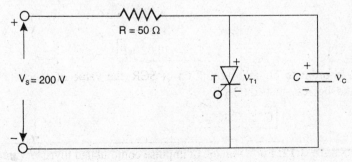

Fig. P. 3.11

Solution: Under steady state condition, current through

$$\text{SCR} = \frac{V_S}{R} = \frac{200}{50} = 4 \text{ Amp and}$$

Voltage across ideal SCR = Voltage V_C across capacitor $C = 0$.

When SCR is forced commutated, capacitor C begins charging from source voltage V_S through resistance R so that capacitor voltage $V_C \, (= V_T)$ is given by

$$V_C = V_S \left[1 - e^{-\frac{t}{Rc}} \right]$$

or
$$\frac{dv_C}{dt} = V_S \cdot e^{-\frac{t}{Rc}} \frac{1}{RC}$$

$$\boxed{\left. \frac{dv_C}{dt} \right|_{t=0} = \frac{V_S}{RC}}$$

The rate of rise of capacitor voltage V_C across SCR may be large.

in case SCR charging current $C_J \left. \dfrac{dv_C}{dt} \right|_{t=0}$ happens to be equal 5 mA. SCR will get turned on. Here C_J is the junction capacitance of SCR.

or
$$\boxed{\left. C_J \cdot \frac{dv_C}{dt} \right|_{t=0} = 5 \text{ mA}}$$

or
$$C_J \cdot \frac{V_S}{R_C} \doteq 5 \times 10^{-3} \text{ Amp}$$

or
$$25 \times 10^{-12} \times \frac{200}{50 \times C} = 5 \times 10^{-3}$$

or
$$C = \frac{20 \times 10^{-12}}{10^{-3}}$$

or
$$\boxed{C = 0.02 \ \mu\text{F}}$$

In order to obviate turning on of SCR, the value of capcitance C should be less than 0.02 μF, i.e.

$$\boxed{C \leq 0.02 \ \mu\text{F}}$$

Problem 3.12: For a voltage or impulse commutated thyristor circuit shown in Fig. P. 3.12. Capacitor is initially charged to V_S with polarity

as shown. Find the circuit turn-off time for the main thyristor in case
$C = 20\ \mu F$, $R = 10\ \Omega$ and $V_S = 230$ V d.c.

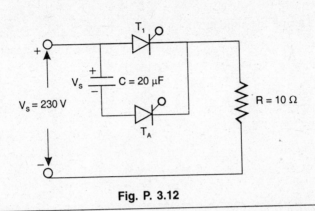

Fig. P. 3.12

Solution: When auxiliary thyristor T_A is turned on, main thyristor T_1
is turned on by means of capacitor voltage V_S appearing as reverse bias.
After T_1 is off, KVL for the circuit consisting of V_S, C, T_A, and R in
series is given by

$$R \cdot i(t) + \frac{1}{C} \int i(t)\,dt = V_S$$

Its laplace transforme is

$$R \cdot I(s) + \frac{1}{C}\left[\frac{I(s)}{S} - \frac{C \cdot V_S}{S}\right] = \frac{V_S}{S}$$

or

$$I(s)\left[R + \frac{1}{SC}\right] = \frac{2V_S}{S}$$

or

$$I(s) = \frac{2V_S}{S} \cdot \frac{SC}{(1+RCS)} = \frac{2V_S}{S} \times \frac{1}{S + \frac{1}{RC}}$$

Its laplace inverse is

$$\boxed{i(t) = \frac{2V_S}{R} \cdot e^{-\frac{t}{RC}}}$$

The voltage across capacitor C is

$$V_C(t) = \frac{1}{C} \int i(t)\,dt + \text{initial voltage across capacitor.}$$

$$= \frac{1}{C} \int \frac{2V_S}{R} \cdot e^{-\frac{t}{RC}} + V_S = V_S \left[1 - 2e^{-\frac{t}{RC}} \right]$$

During the time auxiliary SCR T_A is on.

$$V_C = V_{T1} = V_S \left[1 - e^{-\frac{t}{RC}} \right]$$

The circuit turn off time for T_1 is the time taken by $V_C = V_{T1}$ to change from its value $-V_S$ to zero

$$\therefore \qquad 0 = V_S \left[1 - 2e^{-\frac{t}{RC}} \right]$$

or $\qquad t_C = R_C \ln (2)$
$\qquad\qquad = 10 \times 20 \times 10^{-6} \ln (2) = 138.62 \times 10^{-6}$ sec

$$\boxed{t_C = 138.62 \ \mu \ sec}$$

CHAPTER 4

PHASE-CONTROLLED RECTIFIERS

In phase controlled rectifier their is no need of commutation circuit. In study of thyristor system SCR and diodes are assumed ideal switch, i.e.

(a) There is no voltage drop across them.

(b) No reverse current exists under reverse voltage condition.

(c) Holding current should be zero.

Principle of phase control : An SCR can conducts only when anode voltage is +ve and a gating signal is applied. At the instant of delay angle α, V_o rises from zero to $V_m \sin \alpha$. For resistive load, current i_o is in phase with V_o.

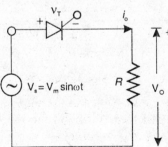

Fig. 4.1

The turn off time of circuit is given as

$$t_C = \frac{\pi}{\omega} \ \sec$$

where $\omega = 2\pi f$

The circuit turn-off time t_C must be more than SCR turn-off time t_q.

$$\therefore \qquad \boxed{t_q = \frac{\pi - \alpha}{\omega} \sec}$$

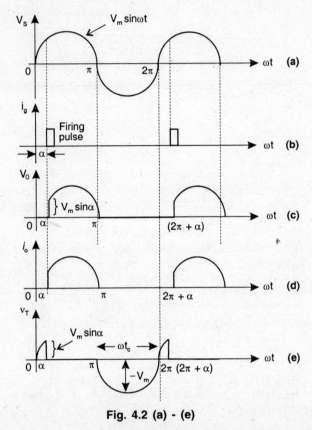

Fig. 4.2 (a) - (e)

Firing angle : It is defined as the angle measured from the instant that gives the largest load voltage to the instant when it is triggered.

Average voltage (V_o) across R

$$V_o = \frac{1}{2\pi} \int_{\alpha}^{\pi} V_m \sin \omega t \, d \, (\omega t)$$

$$\boxed{V_o = \frac{V_m}{2\pi} \, (1 + \cos \alpha)}$$

$$I_o = \frac{V_m}{2\pi R}(1 + \cos \alpha)$$

r.m.s. voltage ($V_{r.m.s.}$) :

$$V_{r.m.s.} = \left[\frac{1}{2\pi}\int_{\alpha}^{\pi}V_m^2 \sin^2 \omega t \cdot d(\omega t)\right]^{\frac{1}{2}}$$

∴

$$V_{r.m.s.} = \frac{V_m}{2\sqrt{\pi}}\left[(\pi - \alpha) + \frac{1}{2}(\sin 2\alpha)\right]^{\frac{1}{2}}$$

$$I_{r.m.s.} = \frac{V_{r.m.s.}}{R}$$

Power deliver to resistive load

$$P_o = V_{r.m.s.} \cdot I_{r.m.s.}$$

∴

$$P_o = I_{r.m.s.} \cdot R$$

Input volt ampear

$$P_i = V_S \cdot I_{r.m.s.}$$

Input power factor $(P \cdot t) = \frac{P_0}{P_i}$

$$P \cdot f = \frac{V_{r.m.s.}}{V_S}$$

Half wave rectifier with

RL Load : Due to presence of inductor L, the load current i_o rises gradually first then decreases.

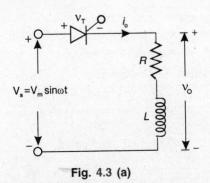

Fig. 4.3 (a)

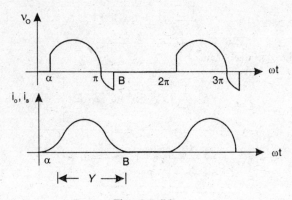

Fig. 4.3 (b)

B = extinction angle and $(\beta - \alpha) = \gamma$ is called conduction angle

$$t_C = \frac{2\pi - B}{\omega}\sec$$

$$t_g = \frac{\beta - \alpha}{\omega}\sec$$

t_C = circuit turn-off time.

t_g = thyristor turn-off time

$$V_o \text{ (average)} = \frac{1}{2\pi}\int_\alpha^\beta V_m \sin \omega t \, d(\omega t)$$

$$V_o = \frac{V_m}{2\pi}(\cos \alpha - \cos \beta)$$

$$V_{\text{r.m.s.}} = \left[\frac{1}{2\pi}\int_\alpha^\beta V_m^2 \sin^2 \omega t \, d(\omega t)\right]^{\frac{1}{2}}$$

$$V_{\text{r.m.s.}} = \frac{V_m}{2\sqrt{\pi}}\left[(\beta - \alpha) - \frac{1}{2}(\sin 2\beta - \sin 2\alpha)\right]^{\frac{1}{2}}$$

RL load with free wheeling diode : The wave of the load current i_o can be improved by connecting a free wheeling (or fly wheeling) diode across the load. With use of free wheeling diode (F.D.) :

(a) Input power factor is improved.

(b) Load current wave form is improved.

(c) Load performance also improved.

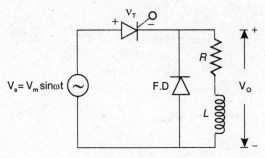

Fig. 4.4

$$V_o = \frac{1}{2\pi} \int_{\alpha}^{\pi} V_m \sin \omega t \, d(\omega t)$$

$$V_o = \frac{V_m}{2\pi} (1 + \cos \alpha)$$

RLE Load

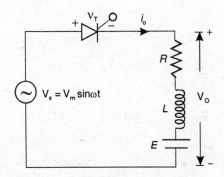

Fig. 4.5

E = battery or counter e.m.f.

The minimum value of firing angle is obtained from the relation

$$\theta_1 = \sin^{-1} (E/V_m)$$

The maximum value of firing angle is obtained from

$$\theta_2 = \pi - \theta_1$$

or

$$\theta_2 = \pi - \sin^{-1} (E/V_m)$$

Peak inverse voltage across thyristor is equal to :

$$PIV = V_m + E$$

$1 - \phi$ half wave converter introduces a d.c. component into the supply line. This is undesirable as it leads to saturation of the supply transformer and other difficulties like harmonics.

This can be over come in $1 - \phi$ full wave converter.

$1 - \phi$ full wave converter with

$1 - \phi$ full wave mid point converter :

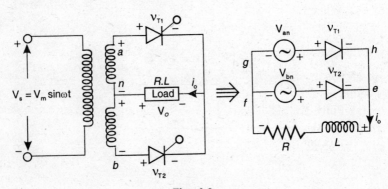

Fig. 4.6

$$V_{an} = + V_{nb} = - V_{bn}$$

$$V_{an} = V_m \sin \omega t$$

$$V_{bn} = - V_m \sin \omega t$$

$$\therefore \qquad V_{ab} = V_{an} + V_{bn} = 2 V_m \sin \omega t$$

When T_1 is conducting $V_{T_1} = 0$, therefore voltage across T_2 at the instant $\omega t = \alpha$ is given by :

$$V_{T_2} = V_{bn} - V_{an} = - 2 V_m \sin \omega t$$

or

$$V_{T_2} = -2 V_m \sin \alpha$$

$$t_{C_2} = \frac{\pi - \alpha}{\omega} \text{ sec} \qquad t_{C_1} = \frac{2\pi - (\alpha + \pi)}{\omega} = \frac{\pi - \alpha}{\omega} \text{ sec}$$

Average voltage

$$V_o = \frac{V_m}{\pi} \int_{\alpha}^{\alpha+\pi} \sin \omega t \cdot d\,(\omega t)$$

$$V_o = \frac{2V_m}{\pi} \cos \alpha$$

We obeserve that

(a) Circuit turn-off time must be great than SCR turn-off time.
(b) When incoming SCR is gated on current is transferred from outgoing SCR to incoming SCR.
(c) When commutation of an SCR is desired it must be reversed biased and the incoming SCR must be forward biased.

1 – φ full wave bridge converter

It is of two types (a) Fully controlled converter (b) Half controlled converter.

(a) Fully controlled converter or two pulse converter uses four thyristors in the form of bridge and is a two quadrant converter.

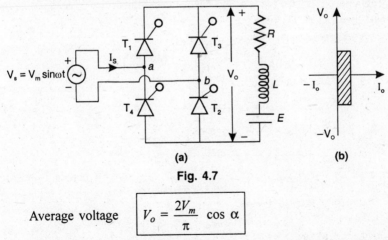

(a)   **(b)**

Fig. 4.7

Average voltage $\qquad \boxed{V_o = \dfrac{2V_m}{\pi} \cos \alpha}$

During α to π : V_o (+ ve), i_o (+ ve) $\therefore$ Power flows from source to load.

During π to $(\pi + \alpha)$: V_o (– ve), i_o (+ ve) $\therefore$ Some power flows from load to source but net power flows from a.c. source to d.c. load because $(\pi - \alpha) > \alpha$.

Average load $V_o = 0$ at $\alpha = 90°$ and $-$ve at $\alpha > 90°$.

The full converter with firing angle delay greater than 90° is called line commutation inverter.

(b) $1 - \phi$ semiconverter :

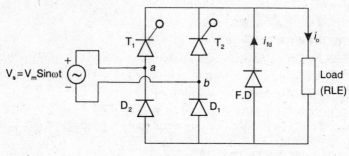

Fig. 4.8

It contains two thyristor and three diode. One diode is connected as free wheel diode. The load current is assumed to be continuous over work range. During the F·D period (π) to $(\pi + \alpha)$, energy stored inductor is recovered and is partially dissipated in R and partially added to the energy stored in load e.m.f.F. No energy is feedback to the source during free wheeling (F·D) period.

Average output voltage

$$V_o = \frac{1}{\pi} \int_\alpha^\pi V_m \sin \omega t \, d(\omega t)$$

$$\boxed{V_o = \frac{V_m}{\pi} (1 + \cos \alpha)}$$

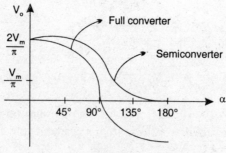

Fig. 4.9

3 – φ thyristor converter

For large power d.c. loads, 3 – φ a.c. to d.c. converters are commonly used. Various types of 3 – φ controlled converters are 3 – φ half wave converter, 3 – φ semiconverter, 3 – φ full converter and 3 – φ dual converter. 3 – φ half wave are rarely used in industry because it introduces d.c. component in the supply current. Semi converter and full converter are quite common in industries. A dual converter is used only when reversible d.c. drives with power rating of several MW are required.

Advantage of 3 – φ converter over 1 – φ converter

(a) In 3 – φ converter, the ripple frequency of the converter output voltage is higher than 1 – φ. So filtering requirement for smoothing out the load current is less.

(b) The load current is mostly continuous in 3 – φ converter. So load performance is better in 3 – φ than 1 – φ.

3 – φ full converter

1, 3, 5 — + ve group
4, 6, 2 — – ve group

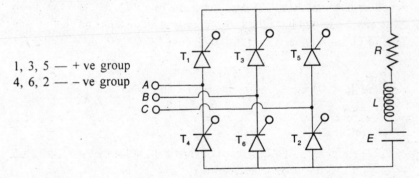

Fig. 4.10

For α = 0° T_1, T_2, T_3, T_4, T_5, T_6 behaves like diode + ve group of SCR are fired at an interval of 120°. Similarly – ve group of SCR are fired at an interval of 120°. But SCR from both the group are fired at an interval of 60°.

$$V_o = \frac{3}{\pi} \int_{-\left(\frac{\pi}{6}-\alpha\right)}^{\left(\frac{\pi}{6}+\alpha\right)} V_m \cos \omega t \cdot d(\omega t)$$

$$\therefore \qquad \boxed{V_o = \frac{3V_m}{\pi} \cos \alpha}$$

3 – ϕ semiconverter

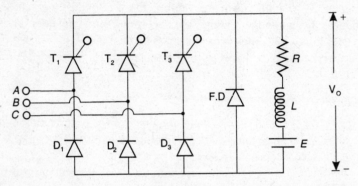

Fig. 4.11

The diode D_1, D_2, D_3, provides a return path for current to the most – ve line terminal. The F·D does not come into play for $\alpha = 15^\circ$. Each SCR and diode conducts for 120°. A 3 – ϕ semiconverter has the unique feature of working as a six pulse converter for $\alpha < 60^\circ$ and 3-pulse for $(\alpha \geq 60^\circ)$.

For $\alpha < 60^\circ$ $V_o = \dfrac{3V_{mp}}{2\pi}(1 + \cos \alpha)$ and for $\alpha \geq 60^\circ$ $V_o = \dfrac{3V_{mp}}{2\pi}$ $(1 + \cos \alpha)$.

Dual converters

Semiconverter are single quadrant converters. In full converters, direction of current can not reverse because of the unidirectional properties of SCR. A full converter operates as a rectifier in first quadrant from $\alpha = 0^\circ$ to 90° and as an inverter from $\alpha = 90^\circ$ to 180° in the fourth quadrant. In first quadrant power flows from a.c. to d.c. load and in fourth quadrant power flows from d.c. load to a.c. source.

In case of dual converter two full converter can be connected back to back to the load circuit.

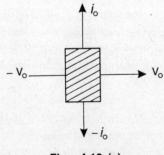

Fig. 4.12 (a)

1 – ϕ dual converter.
3 – ϕ have 12 thyristor.

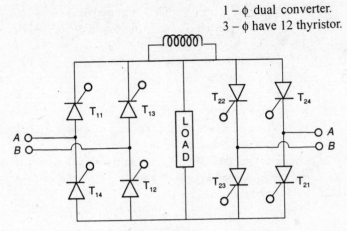

Fig. 4.12 (b)

There are two functional mode of a dual converter. One is non-circulating current mode and other is circulating current mode. Non-circulating type of dual onverter uses 1 – ϕ and 3 – ϕ. Reactor is used to limit the circulating current.

$$\boxed{\text{Firing angle} = \alpha_1 + \alpha_2 = 180^\circ}$$

Average voltage

$$\boxed{V_o = \frac{V_{o1} + V_{o2}}{2}}$$

Point to be noted

(1) For continuous current:
   (a) 1 – ϕ full converter — each pair of SCR conducts for π; free wheel diode for π.
   (b) 1 – ϕ semi converter — ($\pi - \alpha$) for SCR ; 'α' for F·D.
   For discontinuous current :
   (a) 1 – ϕ full converter — ($\beta - \alpha$) for SCR ; o for F·D
   (b) 1 – ϕ semi converter — It $\beta > \pi$, ($\pi - \alpha$) for SCR; ($\beta - \pi$) for F·D.
       If $\beta < \pi$, ($\beta - \alpha$) for SCR; 0 for F·D.

(2) In 1 – ϕ full converter with resistive load and for a firing angle α, the load current is zero and non zero respectively for α, $\pi - \alpha$. For half converter it is at α and π.

(3) With resistive load no F·D is required.

(4) $3 - \phi$ semiconverted, for a firing angle $\leq 60^\circ$, F·D conducts for 0°.

(5) $3 - \phi$ semiconverter, for a firing angle equal to 90° and for continuous conduction each SCR and diode conduct, respectively for 60° and 60°.

(6) $3 - \phi$ semiconverter, for firing angle equal to 120° and B = 110°, each SCR and diode conducts respectively for 60° and 60° and F·D conducts for 50°.

(7) Frequency of the ripple in the output voltage of a $3 - \phi$ semi converter depends on (1) firing angle (2) supply frequency.

(8) In $1 - \phi$ full converter, the number of SCR conducts during overlap is 4 and output voltage during overlap is zero.

PROBLEMS

Problem 4.1: A resistive load of 10 Ω is connected through a half-wave SCR circuit to 220 V, 50 Hz, $1 - \phi$ source. Calculate the power delivered to load for a firing angle of 60°. Find also the value of input power factor.

Solution: r.m.s. voltage at firing angle 60° is equal to :

$$V_{or} = \frac{\sqrt{2} \times 220}{2\sqrt{\pi}} \left(\left(\pi - \frac{\pi}{3} \right) + \frac{1}{2} \sin 2 \times \frac{\pi}{3} \right)^{\frac{1}{2}}$$

$$= 87.76 \left(2.094 + \frac{0.866}{2} \right)^{\frac{1}{2}} = 87.76 \times 1.589$$

$$\boxed{V_{or} = 139.5 \text{ V}}$$

$\therefore$ Power delivered to load ($R = 10 \ \Omega$) is equal to :

$$P = \frac{V_{or}^2}{R} = 1946.886 \text{ W}$$

$\therefore$ $\boxed{P = 1946.886 \text{ W}}$

$\therefore$ Input power factor $= \dfrac{V_{or}}{V_S}$

$$P \cdot f = \frac{139.5}{220}$$

or, $\boxed{P \cdot f = 0.06340}$

Problem 4.2: An RL load, energised from $1 - \phi$, 230 V, 50 Hz source through a single thyristor, has $R = 10 \, \Omega$ and $L = 0.08$ H. If thyristor is triggered in every positive half cycle at $\alpha = 75°$, find current expression as a function of time.

Solution: The circuit diagram of $1 - \phi$ half wave with RL load is :

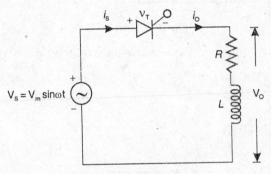

Fig. P. 4.2 (a)

The voltage equation for the circuit is :

$$V_m \sin \omega t = Ri_o + L \frac{di_o}{dt}$$

The load current I_o consists of two component, one steady-state component i_s' and the other transient component i_t'. Here i_s' is given by

$$i_s' = \frac{V_m}{\sqrt{R^2 + (\omega L)^2}} \sin (\omega t - \phi)$$

where $\phi = \tan^{-1} \frac{\omega L}{R}$

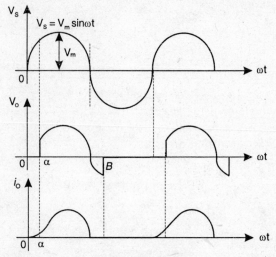

Fig. P 4.2 (b)

The transient component i_t can be obtained from force-free equation:

$$Ri_t' + L\frac{di_t'}{dt} = 0$$

Its solution gives, $i_t' = A e^{-\left(\frac{R}{L}\right)t}$

$\therefore \qquad\qquad i_o = i_s' + i_t'$

$$\boxed{i_o = \frac{V_m}{Z}\sin(\omega t - \phi) + A e^{-\left(\frac{R}{L}\right)t}} \qquad (1)$$

at $\omega t = \alpha$, $i_o = 0$, Thus,

$$0 = \frac{V_m}{Z}\sin(\alpha - \phi) + A e^{-\frac{R\alpha}{\omega L}}$$

or $\qquad\qquad \boxed{A = -\frac{V_m}{Z}\sin(\alpha - \phi)\, e^{\frac{R\alpha}{\omega L}}}$

Putting the value of A in eq. (1), we get

$$\boxed{i_o = \frac{V_m}{Z}\sin(\omega t - \phi) + \left(-\frac{V_m}{Z}\sin(\alpha - \phi)\, e^{-\frac{R\alpha}{\omega L}}\right)e^{-\frac{R}{L}t}}$$

$$\phi = \tan^{-1} \frac{\omega L}{R} = \tan^{-1} \frac{25.1}{10} = 68.30^\circ$$

$$\alpha = 75^\circ, \ Z = \sqrt{R^2 + (\omega L)^2} = 27.04 \ \Omega$$

$$\therefore \ i_o = \frac{\sqrt{2} \times 230}{27.04} \sin(\omega t - 68.30^\circ) - \frac{\sqrt{2} \times 230}{27.04}$$

$$\sin(75^\circ - 68.30) \ e^{-125t} \times e^{\frac{R\alpha}{\omega L}}$$

$$= 12.023 \sin(\omega t - 68.30^\circ) - (12.023 \times 0.1166) \ e^{\frac{R\alpha}{\omega L}} \cdot e^{-125t}$$

$$= 12.023 \sin(\omega t - 68.30^\circ) - 1.40 \ e^{\frac{10 \times 75\pi}{2 \times \pi \times 50 \times 0.08 \times 180}} e^{-125t}$$

$$= 12.023 \sin(\omega t - 68.30^\circ) - 1.40 \times 1.68 \ e^{-125t}$$

$$\boxed{i_o = 12.023 \sin(\omega t - 68.30^\circ) - 2.355 \ e^{-125t}}$$

Problem 4.3: A battery is charged by a $1 - \phi$ one pulse thyristor controlled rectifier. The supply is 30 V, 50 Hz and battery e.m.f., is constant at 6 V. Find the resistance to be inserted in series with the battery to limit the charging current to 4 A on the assumption that SCR is triggered continuously. Take a voltage drop of 1 V across the SCR. Drive the expression used.

Solution:

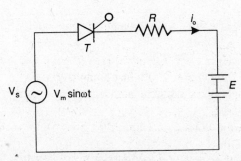

Fig. P. 4.3

For the circuit voltage equation is :
$$V_m \sin \omega t = E + i_o R + 1 \text{ V}$$

or, $$i_o = \frac{V_m \sin \omega t - (E+1)}{R}$$

Since the SCR is turn on when $V_m \sin \theta_1 = E$ and is turned off when $V_m \sin \theta_2 = E$, where $\theta_2 = \pi - \theta_1$.

$$\therefore \qquad \theta_1 = \sin^{-1}\left(\frac{E}{V_m}\right) = \sin^{-1}\left(\frac{6}{30}\right) = 11.53^\circ$$

The battery charging requires only the average current I_o given by:

$$I_0 = \frac{1}{2\pi R}\left[\int_{\theta_1}^{\pi-\theta_1} (V_m \sin \omega t - (E+1)\, d\,(\omega t)\right]$$

$$\boxed{I_0 = \frac{1}{2\pi R}\left[2 V_m \cos \theta_1 - (E+1)(\pi - 2\theta_1)\right]}$$

$$\therefore 4 \text{ Amp} = \frac{1}{2\pi R}\left[2 \times \sqrt{2} \times 30 \cdot \cos 11.53 - (6+1)\left(\pi - \frac{2 \times 11.53\,\pi}{180}\right)\right]$$

$$4 \text{ Amp} = \frac{1}{2\pi R}[83.13 - 19.172]$$

$$4 \text{ Amp} = \frac{1}{2\pi R}[63.95]$$

or $$\boxed{R = \frac{1}{2\pi \times 4}[63.95]\,\Omega}$$

or $$\boxed{R = 2.544\ \Omega}$$

Problem 4.4: A single-phase one-pulse converter with RLE load has the following data :

Supply voltage = 230 V at 50 Hz, $R = 2\,\Omega$, $L = $ mH, $E = 120$ V, extinction angle $B = 220^\circ$, firing angle $\alpha = 25^\circ$.

(1) Calculate the voltage across thyristor at the instant SCR is triggered.
(2) Find the voltage that appears across SCR when current decays to zero.
(3) Find the peak inverse voltage for the SCR.

Solution: In $1 - \phi$ one pulse converter with RLE load the voltage across thyristor is given as :

(1) at the instant SCR is triggered :
$$V_T = V_m \sin \alpha - E$$

So, $\quad V_T = \left(\sqrt{2} \times 230 \cdot \sin 25^\circ - 120\right) = (137.46 - 120)$ Volt

$$\boxed{V_T = 17.46 \text{ V}}$$

(2) at the instant current decay to zero :
$$V_T = E + V_m \sin (B - \pi)$$
$$= 120 + \sqrt{2} \times 230 \sin (220^\circ - 180^\circ)$$
$$= 120 + \sqrt{2} \times 230 \sin 40^\circ = 120 + 209.07$$

$$\boxed{V_T = 339.07 \text{ V}}$$

(3) *PIV* is given as :
$$PIV = V_m + \text{E} = \sqrt{2} \times 230 + 120 = 325.26 + 120$$

$$\boxed{PIV = 445.26 \text{ V}}$$

Problem 4.5: In a $1 - \phi$ mid point converter, turn, ratio from primary to each secondary is 1.25. The source voltage is 230 V, 50 Hz. For a resistive load of $R = 2 \Omega$, determine:
(a) Maximum value of average output voltage and load current and the corresponding firing and conduction angles.
(b) Maximum average and r.m.s. thyristor current.
(c) Maximum possible values of +ve and –ve voltage across SCR.
(d) The value of α for load voltage of 100 V.
(e) The value of voltage across SCR at the instant of commutation for α of part (d).

Solution: Turn ratio of centre tap transformer is 1.25. So, secondary voltage is equal to, $\dfrac{230 \text{ V}}{1.25} = 184$ V

Average value of output voltage is given by :

$$V_{ave} = \frac{1}{\pi} \int_{\alpha}^{\alpha + \pi} V_m \sin \omega t \cdot d\,(\omega t) = \frac{2 V_m}{\pi} \cos \alpha$$

∴ (a) Maximum value of average output Voltage occurs when firing angle α is $0°$

So, $(V_{ave})_{max} = \dfrac{2V_m}{\pi} \cos 0° = \dfrac{2V_m}{\pi} = \dfrac{2 \times \sqrt{2} \times 184}{\pi}$

$$\boxed{(V_{ave})_{max} = 165.65 \text{ V}}$$

$(I_{ave})_{max} = \dfrac{V_{ave}}{R} = \dfrac{165.65}{2\,\Omega}$

$$\boxed{(I_{ave})_{max} = 82.82 \text{ Amp}}$$

Conduction angle $\gamma = 180° - \alpha$

or $\gamma = 180° - 0°$

or $\boxed{\gamma = 180°}$

(b) Maximum average thyristor current

$$= \dfrac{(I_{ave})_{max}}{2}, \text{i.e.} \dfrac{\text{load current}}{2} = \dfrac{82.82}{2} = 41.41 \text{ A}$$

So, $\boxed{(I_{ave})_{max} \text{ through thyristor} = 41.41 \text{ A}}$

and $V_{r.m.s.} = \left[\dfrac{1}{\pi} \int\limits_{\alpha}^{\alpha+\pi} V_m^2 \sin^2 \omega t \, d(\omega t) \right]^{\frac{1}{2}}$

$$V_{r.m.s.} = \dfrac{V_m}{\sqrt{2\pi}} \left[(\alpha+\pi) - \sin\dfrac{2(\alpha+\pi)}{2} - \alpha + \dfrac{\sin 2\alpha}{2} \right]^{\frac{1}{2}}$$

or $(V_{r.m.s.})_{max}$ occurs at $\alpha = 0°$

So, $(V_{r.m.s.})_{max} = \dfrac{V_m}{\sqrt{2\pi}} \left(\sqrt{\pi} \right)$

$$\boxed{(V_{r.m.s.})_{max} = \dfrac{V_m}{\sqrt{2\pi}}}$$

∴ $(I_{r.m.s.})_{max} = \dfrac{V_m}{\sqrt{2\pi}R} = \dfrac{\sqrt{2} \times 184}{\sqrt{2} \times 2} = 92$

So, maximum r.m.s. current through each SCR is equal to :

$$= \frac{92}{2} = 46 \text{ Amp}$$

(c) Maximum possible values of + ve and − ve voltages across SCR is equal to peak inverse voltage.

So, $\qquad PIV = 2 V_m = 2 \times \sqrt{2} \times 184 \text{ V}$

$$\boxed{PIV = 520.43 \text{ V}}$$

(d) For load voltage = 100 V, firing angle α is equal to :

$$100 \text{ V} = \frac{2 V_m}{\pi} \cos \alpha$$

$$100 = \frac{2 \times \sqrt{2} \times 184}{\pi} \cos \alpha$$

or $\qquad \cos \alpha = 0.603$

or $\qquad \alpha = \cos^{-1} (0.603)$

or $\qquad \boxed{\alpha = 52.868^{\circ}}$

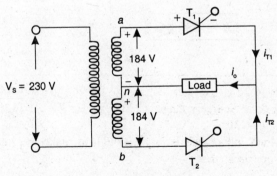

Fig. P. 4.5

(e) The value of voltage across SCR at the instant of commutation for $\alpha = 52.862^{\circ}$ is $= 2 V_m \sin \alpha = 2 \times \sqrt{2} \times 184 \times \sin 52.862^{\circ}$

$$\boxed{= 414.87 \text{ V}}$$

Problem 4.6: Find the value of R in case battery charging current is 6 A, supply voltage is 40 V, 50 Hz, $E = 12$ V. and $V_o = 1$ Volt

Solution: SCR is turn-on when $\theta_1 = \sin^{-1}\left(\dfrac{E}{V}\right)$

So, $\theta_1 = \sin^{-1}\left(\dfrac{12}{40}\right) = 17.45^\circ$

$\therefore$ Average current required for $1 - \phi$ full converter's charging battery is :

$$I_o = \frac{1}{\pi R}\left[2 V_m \cos\theta_1 - (E+1)(\pi - 2\theta_1)\right]$$

or $$6 = \frac{1}{\pi R}\left[2 \times \sqrt{2} \times 40 \cdot \cos 17.45^\circ - (12+1)(2.532)\right]$$

$$R = \frac{1}{\pi \times 6}\left[107.93 - 32.91\right]$$

$$R = \frac{75.01}{\pi \times 6}\ \Omega$$

or $$\boxed{R = 3.978\ \Omega}$$

Problem 4.7: $1 - \phi$ full converter feeding RLE load has the following data. Source voltage = 230 V, 50 Hz $R = 2.5\ \Omega$, $E = 100$ V, firing angle = 30°.

It the load inductance is large enough to make the load current virtually constant, then

(a) Sketch the time variations of source voltage, source voltage, source current, load voltage, load current, current through one SCR and voltage across it.

(b) Compute the average value of load voltage and load current.

(c) Compute the input *p.f.*

Solution: (a)

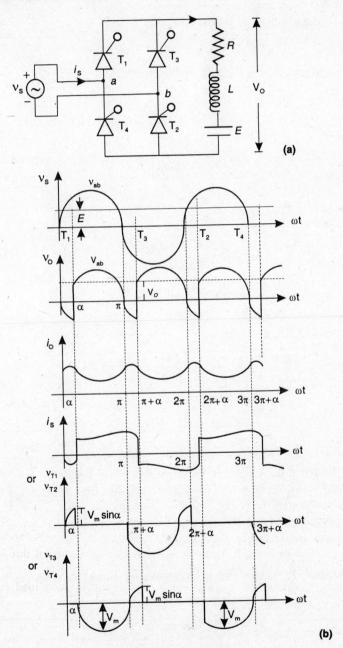

Fig. P. 4.7 (a) - (b)

Voltage, and current wave form for continuous load current.

(b) Average value of load current and voltage is :

$$V_o = \frac{1}{\pi} \int_{\alpha}^{\pi+\alpha} V_m \sin \omega t + dt = \frac{2 V_m}{\pi} \cos \alpha$$

$$= \frac{2 \times \sqrt{2} \times 230}{\pi} \cos 30^\circ$$

$$\boxed{V_o = 179.33 \text{ Volt}}$$

So, $$I_o = \frac{V_o - E}{R}$$

$$I_o = \frac{179.3 - 100}{2.5}$$

or $$\boxed{I_0 = 31.72 \text{ Amp}}$$

(c) Input *P.f* is given as

$$= \frac{V_o}{V_S} = \frac{179.3}{230} = 0.7795$$

So, $$\boxed{P.f = 0.7795}$$

Problem 4.8: A separately-excited d.c. motor fed through a single-phase semiconverter runs at a speed of 1200 rpm when a.c. supply voltage is 230 V, 50 Hz, and the motor counter e.m.f. is 140 V. The firing angle delay is 50°. Armature circuit resistance is 3 Ω. Compute the average armature current and motor torque.

Solution: In single phase semiconverter output average voltage is given as :

$$V_o = \frac{V_m}{\pi} (1 + \cos \alpha)$$

$$V_o = \frac{\sqrt{2} \times 230}{\pi} (1 + \cos 50^\circ)$$

$$\boxed{V_o = 170 \text{ V}}$$

So, terminal voltage (V_t) across motor is equal to V_o

i.e. $\boxed{V_t = V_o = 170 \text{ V}}$

In d.c. motor back e.m.f. (E_b) is equal to :

$$E_b = V_t - I_a r_a$$

or $140 \text{ V} = 170 - I_a \times 3$

or $3 I_a = 170 - 140$

or $I_a = \dfrac{30}{3} = 10 \text{ Amp}$

So, $\boxed{I_a = 10 \text{ Amp}}$

Now, motor torque is given as :

$$E_b I_a = T \omega_s$$

$$\omega_s = \frac{2 \pi Ns}{60} = \frac{2 \pi \times 1200}{60} = 40 \pi$$

$\therefore$ $T = \dfrac{E_b I_a}{\omega}$ N.m

or $T = \dfrac{140 \times 10}{40\pi}$ N.m

or $T = \dfrac{35}{\pi}$ N.m

or $\boxed{T = 11.14 \text{ N.m}}$

Problem 4.9: A single-phase full-converter supplies power to RLE load. The source voltage is 230 V, 50 Hz and for load $R = 2\ \Omega$, $L = 10$ mH, $E = 100$ V. For a firing angle of 30°, find he average value of output current and output voltage is case the load current extinguishes at (a) 200°, and (b) 170°.

Solution: $\qquad V_S = V_m \sin \omega t$

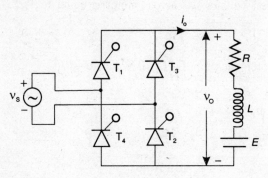

Fig. P. 4.9

$1 - \phi$ full wave converter.

For the circuit, the voltage equation is

$$V_m \sin \omega t = E + i_o R$$

or $\qquad i_o = \dfrac{V_m \sin \omega t - E}{R}$

Since the SCR is turned on when $\alpha = 30°$ and turn-off when $\beta = 200°$ or $170°$. So

$$I_o = \frac{2}{2\pi R} \left[\int_{\alpha}^{\beta} (V_m \sin \omega t - E)\, d\,(\omega t) \right]$$

$$= \frac{2}{2\pi R} \left[(-V_m \cos \omega t) - E(\omega t) \right]_{\alpha}^{\beta}$$

$$= \frac{2}{2\pi R} \left[(V_m (\cos \alpha - \cos \beta) - E(\beta - \alpha)) \right]$$

(a) For $\beta = 200°$:

$$I_o = \frac{2}{2\pi R} = \left[\sqrt{2} \times 230 \, (\cos 30° - \cos 200°) - 100 \, (200 - 30°) \times \frac{\pi}{180} \right]$$

$$= \frac{2}{2\pi R} \left[587.34 - 296.7\right] = \frac{2 \times 290.64}{2\pi \times 2} = 2 \times 23.128 \text{ Amp}$$

or $\boxed{I_o = 2 \times 23.128 \text{ Amp}} = 46.25 \text{ Amp}$

or $\boxed{I_o = 46.25 \text{ Amp}}$

$V_o = I_o R + E$ [as voltage across induction is 0]

$\qquad = 46.25 \times 2 + 100 = 92.51 + 100$

$\boxed{V_o = 192.51 \text{ V}}$

(b) at $\beta = 170°$:

$$I_o = \frac{2}{2\pi R}\left[V_m \left(\cos \alpha - \cos \beta\right) - E\left(\beta - \alpha\right)\right]$$

$$I_o = \frac{1}{\pi \times 2}\left[\sqrt{2} \times 230 \left(\cos 30° - \cos 170°\right) - 100 \left(170° - 30°\right)\frac{\pi}{180}\right]$$

$$= \frac{1}{2\pi}\left[602.01 - 244.346\right] = \frac{357.66}{2\pi} = 56.92$$

$\boxed{I_o = 56.92 \text{ Amp}}$

$\therefore \qquad V_o = I_o R + E = 56.92 \times 2 + 100 = 113.84 + 100$

$\boxed{V_o = 213.84 \text{ V}}$

Problem 4.10: For a $3 - \phi$ half wave diode rectifier, derive an expression for the average output voltage V_o in terms of maximum value of source voltage from line to neutral.

It this rectifier feeds RL load with $R = 5 \ \Omega$ and $L = 3$ mH, find the average load current for $3 - \phi$ input voltage of 400 V, 50 Hz.

Solution: For a $3 - \phi$ diode rectifier each diode conducts for 120° or $\frac{2\pi}{3}$ radian. It is seen that diode A_1 conducts from $\omega t = 30°$ to 150° and diode $B_1 = 150°$ to 270° and diode $C_1 = 270°$ to 390°. So,

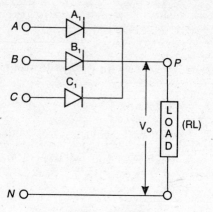

Fig. P. 4.10

average output voltage V_o is given as

$$V_o = \frac{3}{\text{periodic time}} \int_{\alpha_1}^{\alpha_2} V_a \, d\,(\omega t) = \frac{3}{2\pi} \int_{\frac{\pi}{6}}^{\frac{5\pi}{6}} V_{mp} \sin\omega t \, d\,(\omega t)$$

or $\quad V_o = \dfrac{3V_{mp}}{2\pi}\left[\dfrac{\cos\pi}{6} - \dfrac{\cos5\pi}{6}\right] = \dfrac{3\sqrt{3}\,V_{mp}}{2\pi} = \dfrac{3\sqrt{6}\,V_{ph}}{2\pi}$

or $\qquad \boxed{V_o = \dfrac{3\sqrt{6}\,V_{ph}}{2\pi}}$

$$V_{ph} = \frac{400}{\sqrt{3}} = 230.94 \text{ V}$$

$$\therefore \quad V_o = \frac{3\sqrt{6}\times 230.94}{2\pi} = 270.09 \text{ V}$$

$$I_o = \frac{3\sqrt{6}\times 230.94}{2\pi\,R} = 54.01 \text{ Amp}$$

$$\therefore \quad \boxed{I_o = 54.01 \text{ Amp}}$$

Problem 4.11: A $3-\phi$ half wave SCR converter delivers constant load current of 30 A over the firing angle range of 0° to 80°. At these two firing angles, compute the power delivers to load for an a.c. input voltage of 400 V from a delta-star transformer.

Solution: Average output voltage of 3 – φ half wave rectifier are given by :

$$V_o = \frac{3\sqrt{3}}{2\pi} V_{mp} \cos \alpha \text{ for } 0 < \alpha < \frac{\pi}{6}$$

$$= \frac{3}{2\pi} V_{mp} \left(1 + \cos\left(\alpha + \frac{\pi}{6}\right) \right) \text{ for } \frac{\pi}{6} < \alpha < \frac{5\pi}{6}$$

$$V_{ph} = \frac{400}{\sqrt{3}} = 230.94 \text{ V}$$

$$\therefore \qquad V_{mp} = \sqrt{2} \ V_{ph} = 326.59 \text{ V}$$

∴ (1) at firing angle $\alpha = 0^\circ$

$$V_o = \frac{3\sqrt{3}}{2\pi} \times 326.59 \cos 0^\circ$$

$$\boxed{V_o = 270.09 \text{ V}}$$

∴ Power delivered to load is equal to :

$$P_o = I_0 \ V_o = 30 \times 270.09$$
$$P_o = 8102.84 \text{ W}$$

$$\boxed{P_o = 8.102 \text{ kW}}$$

(2) at firing angle $\alpha = 80^\circ$

$$V_o = \frac{3}{2\pi} V_{mp} \left(1 + \cos\left(\alpha + \frac{\pi}{6}\right) \right)$$

$$= \frac{3}{2\pi} \times 326.59 \ (1 + \cos 110^\circ)$$

$$= \frac{3}{2\pi} \times 326.59 \ (1 - 0.342) = \frac{3 \times 326.59}{2\pi} \times 0.657$$

$$\boxed{V_o = 102.6 \text{ Volt}}$$

∴ Power delivered to the load at α = 80° is equal to
$$P_o = I_o V_o = 30 \times 102.6 \text{ W} = 3078 \text{ W}$$

$$\boxed{P_o = 3.078 \text{ kW}}$$

[Note : For firing angle greater than 30°, the output voltage is discontinuous for a resistive load, so the limit of integration should be from $\left(\dfrac{\pi}{6}+\alpha\right)$ to π and $V_o = \dfrac{3 V_{mp}}{2\pi}\left[1+\cos\left(\alpha+\dfrac{\pi}{6}\right)\right]$.

Problem 4.12: (a) A resistive load of 10 Ω is connected to a 3–φ full converter. The load takes 5 kW for a firing angle delay of 70°. Find the magnitude of per phase input supply voltage. Derive the expression required for the output voltage in terms of firing angle etc.
(b) Repeat part (a) in case an inductor connected in series with the load makes the load current constant.
(c) Repeat part (a) in case an inductor connected with the load makes the load current continuous.

Solution: (a) For α = 70°, the output voltage is discontinuous, for a resistive load. So, r.m.s. value of output voltage is :

$$V_{or}^2 = \frac{3}{\pi}\int\limits_{-\left(\frac{\pi}{6}-\alpha\right)}^{\frac{\pi}{2}} V_{ml}^2 \cos^2 \omega t \, d(\omega t)$$

$$= \frac{3}{\pi} V_{ml}^2 \int\limits_{-\left(\frac{\pi}{6}-\alpha\right)}^{\frac{\pi}{2}} \frac{\cos 2\omega t + 1}{2} d(\omega t)$$

$$= \frac{3}{2\pi} V_{ml}^2 \left[\frac{\sin 2\omega t}{2}+(\omega t)\right]_{-\left(\frac{\pi}{6}-\alpha\right)}^{\frac{\pi}{2}}$$

$$= \frac{3 V_{ml}^2}{2\pi}\left[\sin 2\times\frac{\pi}{2} - \frac{\sin 2\left(\frac{\pi}{6}-\alpha\right)}{2} + \frac{\pi}{2}+\frac{\pi}{6}-\alpha\right]$$

or $\quad V_{or}^2 = \dfrac{3 V_{ml}^2}{2\pi}\left[\sin\pi + \dfrac{\sin\left(\dfrac{2\pi}{6}-2\alpha\right)}{2} + \dfrac{2\pi}{3}-\alpha\right]$

$\qquad\quad = \dfrac{3 V_{ml}^2}{2\pi}\left[0 + \left(\dfrac{2\pi}{3}-\alpha\right) + \dfrac{\sin\left(\dfrac{\pi}{3}-2\alpha\right)}{2}\right]$

$\qquad\quad = \dfrac{3 V_{ml}^2}{2\pi}\left[\left(\dfrac{2\pi}{3}-\alpha\right) + \dfrac{1}{2}\left(\sin\dfrac{\pi}{3}\cdot\cos 2\alpha - \dfrac{\cos\pi}{3}\cdot\sin 2\alpha\right)\right]$

$\qquad\quad = \dfrac{3 V_{ml}^2}{2\pi}\left[\left(\dfrac{2\pi}{3}-\alpha\right) + \dfrac{1}{2}\left(\dfrac{\sqrt{3}}{2}\cdot\cos 2\alpha - \dfrac{1}{2}\sin 2\alpha\right)\right]$

$V_{or}^2 = \dfrac{3 V_{ml}^2}{2\pi}\left[\left(\dfrac{2\pi}{3}-\alpha\right) + \dfrac{1}{4}\left(\sqrt{3}\cos 2\alpha - \sin 2\alpha\right)\right]$

$\therefore\quad V_{or} = \dfrac{V_{ml}\sqrt{3}}{\sqrt{2\pi}}\left[\left(\dfrac{2\pi}{3}-\alpha\right) + \dfrac{1}{4}\left(\sqrt{3}\cos 2\alpha - \sin 2\alpha\right)\right]^{\frac{1}{2}}$

or $\quad \boxed{V_{or} = V_{ml}\sqrt{\dfrac{3}{2\pi}}\left[\left(\dfrac{2\pi}{3}-\alpha\right) + \dfrac{1}{4}\left(\sqrt{3}\cos 2\alpha - \sin 2\alpha\right)\right]^{\frac{1}{2}}}$

or $\quad \dfrac{V_{or}^2}{R} = 5000$ watts, and $V_{ml} = \sqrt{2}\, V_S$

$\therefore\qquad \dfrac{3 V_{ml}^2}{2\pi R}\left[\dfrac{5\pi}{18} + \dfrac{1}{4}(-1.326 - 0.642)\right] = 5000$

or $\qquad\qquad \dfrac{3 V_{ml}^2}{2\pi R}\left[\dfrac{5\pi}{18} - 0.492\right] = 5000$

or $\qquad\qquad \dfrac{3 V_{ml}^2}{2\pi \times 10}\left[0.380\right] = 5000$

or $\qquad\qquad\qquad V_{ml}^2 = \dfrac{5000 \times 2\pi \times 10}{3 \times 0.380}$

or $\boxed{V_{ml} = 524.49 \text{ Volt}}$

$$V_S = \frac{V_{ml}}{\sqrt{2}} = 370.87 \text{ V}$$

Per phase voltage $= \frac{V_S}{\sqrt{3}} = 214.24 \text{ V}$

(b) For constant load current, average load current I_o is equal to r.m.s. load current, I_{or} :

$\therefore$ $\qquad\qquad\qquad I_o^2 \, R = 5 \text{ kW}$

or $\qquad\qquad\qquad \dfrac{V_o^2}{R} = 5 \text{ kW}$

or $\qquad \left[\dfrac{3V_{ml}}{\pi}\cos\alpha\right]^2 \times \dfrac{1}{R} = 5000 \text{ W}$

or $\qquad\qquad \left[\dfrac{3V_{ml}}{\pi}\cos\alpha\right] = (50000)^{1/2}$

$$\frac{3V_{ml}\cos\alpha}{\pi} = 223.60$$

or $\qquad\qquad V_{ml} = \dfrac{223.60 \times \pi}{3\cos\alpha}$

or $\qquad\qquad V_{ml} = \dfrac{223.60 \times \pi}{3\cos 70^\circ}$

or $\qquad\qquad \boxed{V_{ml} = 684.63 \text{ V}}$

$\therefore$ $\qquad V_S = \dfrac{V_{ml}}{\sqrt{2}} = \dfrac{684.63}{\sqrt{2}} = 484.11 \text{ V}$

or $\qquad\qquad \boxed{V_S = 484.11 \text{ V}}$

and perphase voltage $V_{ph} = \dfrac{484.11}{\sqrt{3}} \text{ V}$

or $\qquad \boxed{V_{ph} = 279.50 \text{ V}}$

(c) For continuous load current :

$$V_{or} = V_{ml} \sqrt{\frac{3}{2\pi}} \left[\frac{\pi}{3} + \frac{\sqrt{3}}{2} \cos 2\alpha \right]^{\frac{1}{2}}$$

or
$$\frac{V_{or}^2}{R} = 5000 \text{ W}$$

or
$$\frac{3 V_{ml}^2}{2\pi R} \left[\frac{\pi}{3} + \frac{\sqrt{3}}{2} \cos 2\alpha \right] = 5000$$

or
$$\frac{3 V_{ml}^2}{2\pi R} \left[1.047 + (-0.6634) \right] = 5000$$

or
$$\frac{3 V_{ml}^2 \times 0.3835}{2\pi \times 10} = 5000$$

or
$$V_{ml}^2 = \frac{5000 \times 2\pi \times 10}{3 \times 0.3835}$$

or
$$\boxed{V_{ml} = 522.49 \text{ Volt}}$$

$$\therefore \qquad V_S = \frac{V_{ml}}{\sqrt{2}} = 369.46 \text{ V}$$

or phase voltage $V_{ph} = \dfrac{V_S}{\sqrt{3}}$

or
$$\boxed{V_{ph} = 213.30 \text{ Volt}}$$

Problem 4.13: A separately-excited d.c. motor fed from $3 - \phi$ semiconverter develops a full load torque at 1500 r.p.m. when firing angle is zero, the armature taking 50 A at 400 V d.c. having an armature circuit resistance of 0.5 Ω. Calculate the supply voltage per phase. Find also the range of firing angle required to give speeds between 1500 rpm and 750 rpm at full-load torque.

Solution: For firing angle $\alpha = 0°$, average output voltage of $3 - \phi$ semiconverter is equal to :

$$V_o = \frac{3V_{ml}}{2\pi}(1 + \cos \alpha) = \frac{3V_{ml}}{2\pi}(1 + \cos 0^\circ)$$

$$= \frac{3V_{ml}}{2\pi} \times 2 = \frac{3V_{ml}}{2\pi}$$

Given $V_o = 400$ V

$\therefore$ $400 \text{ V} = \frac{3V_{ml}}{\pi}$

or $V_{ml} = \frac{400 \times \pi}{3} = 418.87$ V

$\therefore$ $V_S = \frac{418.87}{\sqrt{2}} = 296.192$ V

Per phase voltage

$$V_{ph} = \frac{V_S}{\sqrt{3}} = 296.192 \text{ V}$$

or $\boxed{V_{ph} = 171.0 \text{ V}}$ àt 1500 rpm.

For a d.c. machine : (back e.m.f.) $E_b \propto \omega_s$ (speed)

$\therefore$ $\frac{E_{b_1}}{E_{b_2}} = \frac{\omega_{S_1}}{\omega_{S_2}}$

or $\frac{V_{t_1} - I_a r_a}{V_{t_2} - I_a r_a} = \frac{1500}{750}$ [as load current is same]

So, $\frac{400 - 50 \times 0.5}{V_{t_2} - 50 \times 0.5} = 2$

or $\frac{375}{2} = V_{t_2} - 25$

or $V_{t_2} = 187.5 + 25$

or $\boxed{V_{t_2} = 212.5 \text{ V}}$

$\because$ $V_o = \frac{3V_{ml}}{2\pi}(1 + \cos \alpha)$

$$\therefore \qquad 212.5 = \frac{3 \times 418.87}{2\pi} (1 + \cos \alpha)$$

$$\therefore \qquad (1 + \cos \alpha) = \frac{212.5 \times 2\pi}{3 \times 418.87}$$

$$(1 + \cos \alpha) = 1.0625$$

or $\qquad \cos \alpha = 0.0625$

or $\qquad \alpha = \cos^{-1} (0.0625)$

$$\boxed{\alpha = 86.415°}$$

So, range of firing angle to give speeds between 1500 r.p.m. to 750

r.p.m. is equal to : $\boxed{0° \text{ to } 86.415°}$

Problem 4.14: A battery is charged from $3 - \phi$ supply mains of 230 V, 50 Hz through a $3 - \phi$ semiconverter. The battery e.m.f. is 190 V and its internal resistance is 0.5 Ω. An inductor connected in series with the battery renders the charging current of 20 A ripple free. Compute the firing angle delay and the supply power factor.

Solution:

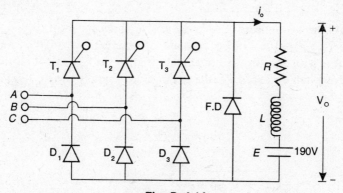

Fig. P. 4.14

output of a $3 - \phi$ semiconverter is equal to :

$$V_o = \frac{3 V_{ml}}{2\pi} (1 + \cos \alpha)$$

Where α is the firing angle.

$$V_o = i_o R + E \text{ [as voltage across inductor is zero]}$$

or
$$V_o = 20 \times 0.5 + 190$$
$$V_o = 10 + 190 = 200 \text{ V}$$

or
$$\boxed{V_o = 200 \text{ V}}$$

$$V_S = 230 \text{ V}$$

$\therefore$
$$V_{ml} = \sqrt{2} \; V_S = \sqrt{2} \times 230 \text{ V}$$

or
$$\boxed{V_{ml} = 325.26 \text{ V}}$$

$\therefore$
$$V_o = \frac{3 V_{ml}}{2\pi} (1 + \cos \alpha)$$

or
$$200 = \frac{3 \times 325.26}{2\pi} (1 + \cos \alpha)$$

or
$$(1 + \cos \alpha) = \frac{200 \times 2\pi}{3 \times 325.26}$$

$$(1 + \cos \alpha) = 1.2877$$

or
$$\cos \alpha = 0.2877$$

or
$$\alpha = \cos^{-1} (0.2877)$$

or
$$\boxed{\alpha = 73.27^\circ}$$

For firing angle $\alpha > 60^\circ$, each SCR conducts for $180^\circ - \alpha$, i.e. $(180 - 73.27^\circ) = 106.73^\circ$. For constant load current of $I_o = 20$ A, supply current i_A is of square wave of amplitude 20 A. As i_A flows for 106.73° over every half cycle of 180°, the r.m.s. value of supply current I_S is given by :

$$I_S = \left[\frac{1}{\pi}(20)^2 \times \frac{106.73 \times \pi}{180} \right]^{\frac{1}{2}} = 20 \sqrt{\frac{106.73}{180}}$$

$$\boxed{I_S = 15.40 \text{ Amp}}$$

Power delivered to load is equal to

$$= E I_o + I_o^2 R = 190 \times 20 + (20)^2 \times 0.5$$
$$= 3800 + 200 = 4000 \text{ W}$$

$$\therefore \text{Input supply } p.f = \frac{4000}{\sqrt{3} \times 230 \times 15.4}$$

$$\boxed{p.f = 0.6519}$$

Problem 4.15: (a) A 3 – φ full converter is used for charging a battery with an e.m.f of 110 V and an internal resistance of 0.2 Ω. For a constant charging current of 10 A, compute the firing angle delay and the supply power factor, for a.c. line voltage of 220 V.
(b)   For the purpose of delivering energy from d.c. source to 3 – φ system, the firing angle of the 3 – φ converter has been increased to 150°. For the same value of d.c. source current of 10 A, compute the output a.c. line voltage of 220 V.

Solution: (a) output voltage $V_o = \dfrac{3 V_{ml}}{\pi} \cos \alpha$

$$\therefore \qquad V_o = I_o R + E$$

or $\qquad \dfrac{3 V_{ml}}{\pi} \cos \alpha = 10 \times 0.2 + 110$

$$\dfrac{3 \times \sqrt{2} \times 220}{\pi} \cos \alpha = 112$$

or $\qquad \cos \alpha = \dfrac{112}{297.104}$

or $\qquad \alpha = \cos^{-1} \left(\dfrac{112}{297.104} \right)$

or $\qquad \boxed{\alpha = 67.85^o}$

For firing angle $\alpha < 60^o$, each SCR conducts for $180 - \alpha$, i.e. $(180^o - 67.85^o) = 112.15^o$. For constant load current of $I_o = 10$ A, supply current i_A is of square wave of amplitude 10 A. As i_A flows for 112.15^o over every half cycle of 180°, the r.m.s. supply current I_S is given as :

$$I_S = 10 \sqrt{\frac{112.15}{180}} = 7.89 \text{ Amp}$$

So, $\boxed{I_S = 7.89 \text{ Amp}}$

Power delivered to load = $V_o I_o = 112 \times 10 = 1120$ W

$\therefore$ Input supply $p.f = \dfrac{1120}{\sqrt{3} \times 220 \times 7.89}$

$\boxed{p.f = 0.372}$

(b) When energy is delivered from d.c. source to $3 - \phi$ system, then the circuit equation is given as :

$$E = I_0 R + V_o$$

or $E = I_0 R + \dfrac{3 V_{ml}}{\pi} \cos \alpha'$

where $\alpha' = 180^\circ - 150^\circ = 30^\circ$

$\therefore$ $110 \text{ V} = 2 \text{ V} + \dfrac{3 V_{ml}}{\pi} \cos \alpha'$

or $108 = \dfrac{3 \times \sqrt{2} \times V_S}{\pi} \cos 30^\circ$

or $V_S = \dfrac{108 \pi}{3\sqrt{2} \cos 30^\circ}$

or $V_S = 92.34 \text{ V}$

Problem 4.16: A single-phase full converter fed from 220 V, 50 Hz supply gives an output voltage of 180 V at no load. When loaded with a constant output current of 10 A, the overlap angle is found to be 6°. Compute the value of source inductance in henries.

Solution:

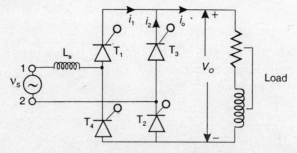

Fig. P. 4.16 (a): $1 - \phi$ full converter with source inductance L_S

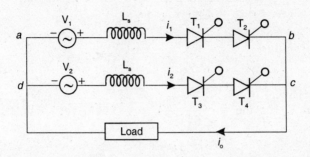

Fig. P. 4.16 (b): Equivalent circuit of full converter with source inductance L_S

Since i_1, increase from 0 to I_0 in the overlap angle μ.

$$\therefore \qquad I_o = \frac{V_m}{L_S} \int_{\frac{\alpha}{\omega}}^{\frac{(\alpha+\mu)}{\omega}} \sin \omega t \cdot dt$$

$$\boxed{I_o = \frac{V_m}{\omega L_S} \left[\cos \alpha - \cos (\alpha + \mu) \right]} \qquad \ldots(1)$$

and output voltage V_o is zero from α to $\alpha + \mu$.

Thus average output voltage V_o is given by

$$V_o = \frac{V_m}{\pi} \int_{(\alpha+\mu)}^{(\alpha+\pi)} \sin \omega t \cdot d(\omega t)$$

$$\boxed{V_o = \frac{V_m}{\pi} \left[\cos \alpha + \cos (\alpha + \mu) \right]} \qquad \ldots(2)$$

at no load, $\mu = 0$. Thus

$$\boxed{V_o = \frac{2 V_m}{\pi} \cos \alpha} \qquad \ldots(3)$$

So, $\qquad 180 \text{ V} = \dfrac{2 \times \sqrt{2} \times 220}{\pi} \cos \alpha$

or $\qquad \cos \alpha = 0.908$

or $\alpha = \cos^{-1}(0.908)$

$$\boxed{\alpha = 24.66°}$$

So, $I_o = \dfrac{V_S}{\omega L_S}\left[\cos\alpha - \cos(\alpha + \mu)\right]$

or   10 Amp $= \dfrac{\sqrt{2} \times 220}{100\pi\ L_S}\left[\cos 24.66° - \cos 124.66° + 6°\right]$

or   10 Amp $= \dfrac{\sqrt{2} \times 220}{100\pi\ L_S}\left[0.908 - 0.860\right]$

or $L_S = \dfrac{\sqrt{2} \times 220}{100\pi \times 10}\left[0.048\right]$

or $L_S = 4.75' \times 10^{-3}$

or $\boxed{L_S = 4.75\ \text{mH}}$

Problem 4.17: (a) Show that the performance of a $1 - \phi$ full converter as effected by source inductance is given by the relation

$$\cos(\alpha + \mu) = \cos\alpha - \frac{\omega L_S\ I_0}{V_m}$$

where the symbols used have their usual meaning.
(b)   A $1 - \phi$ full converter is connected to a.c. supply of $330 \sin 314\ t$

Volt and 50 Hz. It operates with a firing angle $\alpha = \dfrac{\pi}{4}$ rad. The total

load current is maintained constant at 5 A and the load voltage is 140 V. Calculate the source inductance, angle of overlap and the load resistance.

Solution: (a) Equivalent circuit of $1 - \phi$ full converter with source inductance is given as :

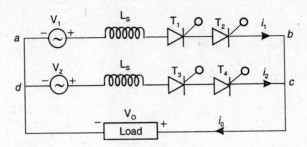

Fig. P. 4.17

and its typcial current and voltage wave form is given as :

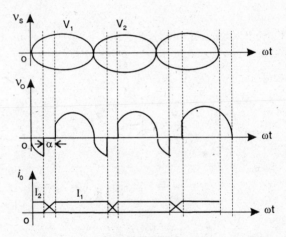

Fig. P. 4.17 (a)

[Note : Here μ = overlap angle]

Now, applying KVL in *abcda* in Fig (a) gives

$$V_1 - L_S \frac{di_1}{dt} = V_2 - L_S \frac{di_2}{dt}$$

or $\quad V_1 - V_2 = L_S \left(\frac{di_1}{dt} - \frac{di_2}{dt} \right)$

It is seen that if $V_1 = V_m \sin \omega t$, then $V_2 = -V_m \sin \omega t$

$$\therefore \ L_S \left(\frac{di_1}{dt} - \frac{di_2}{dt} \right) = 2 V_m \sin \omega t \qquad \qquad ...(1)$$

As the load current is assumed constant

So, $i_1 + i_2 = I_0$

or $\dfrac{di_1}{dt} + \dfrac{di_2}{dt} = 0$ ...(2)

Now, eq. (1) become :

$$\boxed{\dfrac{di_1}{dt} - \dfrac{di_2}{dt} = \dfrac{2V_m}{L_S} \sin \omega t}$$ ...(3)

adding eq. (2) in eq. (3), we get :

$$\boxed{\dfrac{di_1}{dt} = \dfrac{V_m}{L_S} \sin \omega t}$$ ...(4)

Load current i_1 through thyristor pair T_1, T_2, builds up from zero to I_0, during the overlap angle μ, i.e. at $\omega t = \alpha$

$i_1 = 0$ and at $\omega t = (\alpha + \mu)$, $i_1 = I_o$

So, $\displaystyle\int_0^{I_1} di_1 = \dfrac{V_m}{L_S} \int_{\frac{\alpha}{\omega}}^{\frac{(\alpha+\mu)}{\omega}} \sin \omega t \cdot dt$

or $\boxed{I_o = \dfrac{V_m}{\omega L_S} [\cos \alpha - \cos(\alpha + \mu)]}$ ...(5)

or $\cos(\alpha + \mu) = -\dfrac{\omega L_S I_o}{V_m} + \cos \alpha$

or $\boxed{\cos(\alpha + \mu) = \cos \alpha - \dfrac{\omega L_S I_o}{V_m}}$ ...(6)

(b) It is seen that output voltage V_o is zero from α to $(\alpha + \mu)$. Thus average output voltage V_o is given by :

$$V_o = \dfrac{V_m}{\pi} \int_{\alpha+\mu}^{(\alpha+\pi)} \sin \omega t \cdot d(\omega t)$$

$$\boxed{V_o = \dfrac{V_m}{\pi} [\cos \alpha + \cos(\alpha + \mu)]}$$ ...(7)

Since $\cos(\alpha + \mu) = \cos\alpha - \dfrac{\omega L_S I_o}{V_m}$

So, eq. (7) becomes :

$$V_o = \frac{2V_m}{\pi}\cos\alpha - \frac{\omega L_S}{\pi}I_o \qquad \ldots(8)$$

Given $I_o = 5$ A, $\alpha = \dfrac{\pi}{4}$ rad, $V_o = 140$ V

$$V_m = 330 \text{ V} = 330.00 \text{ V}, \ \omega = 100\,\pi$$

So, $\qquad V_o = \dfrac{2V_m}{\pi}\cos\alpha - \dfrac{\omega L_S I_o}{\pi}$

or $\qquad 140 = \dfrac{2 \times 330.00}{\pi}\cos\dfrac{\pi}{4} - \dfrac{100\pi\, L_S \times 5}{\pi}$

$$140 = 148.68 - \frac{500\pi\, L_S}{\pi}$$

or $\qquad \dfrac{500\pi\, L_S}{\pi} = 148.68 - 140$

or $\qquad L_S = \dfrac{8.68 \times \pi}{500\,\pi}$

or $\qquad L_S = \dfrac{8.68}{500}$ H

or $\qquad L_S = 0.01710$ H

or $\qquad \boxed{L_S = 17.10 \text{ mH}}$

$\because \qquad \cos(\alpha + \mu) = \cos\alpha - \dfrac{\omega L_S I_o}{V_m}$

or $\qquad \cos(\alpha + \mu) = \cos 45^\circ - \dfrac{100\pi \times 17.10 \times 10^{-3} \times 5}{330}$

or $\qquad \cos(\alpha + \mu) = 0.625$

or $\qquad (\alpha + \mu) = \cos^{-1}(0.625)$

$$\boxed{(\alpha + \mu) = 51.26^\circ}$$

or $45 + \mu = 51.26^\circ$

or $\mu = 51.26^\circ - 45^\circ$

or $\boxed{\mu = 6.26^\circ}$

So, angle of overlap $= 6.26^\circ$

load resistance $= \dfrac{140 \text{ V}}{5 \text{ A}} = 28 \, \Omega$

$\therefore$ $\boxed{R_L = 28 \, \Omega}$

Problem 4.18: A single phase semiconverter with two thyristor and two diodes as shown in Fig. P. 4.16 is supplied from 230 V, 50 Hz source. The load consists of $R = 10 \, \Omega$, $E = 100$ V and a large inductance to make the load current level. For a firing angle delay of 30°, determine :

(a) Average voltage.

(b) Average output current.

(c) Average and r.m.s. values of thyristor as well as diode current.

(d) Input power factor.

(e) Circuit turn off time.

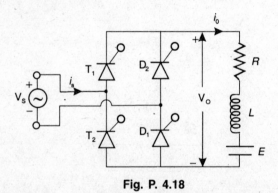

Fig. P. 4.18

Solution: (a) Average output voltage of $1 - \phi$ semiconverter is given as :

$$V_o = \frac{V_m}{\pi}(1 + \cos \alpha) = \frac{\sqrt{2} \times 230}{\pi}(1 + \cos 30^\circ)$$

$$\boxed{V_o = 193.20 \text{ V}}$$

(b) Average output current is equal to :

$$I_o = \frac{V_o - E}{R} = \frac{193.20 - 100}{10} = \frac{93.20}{10}$$

$$\boxed{I_o = 9.32 \text{ Amp}}$$

(c) Average current through thyristor

$$I_{T.A.} = I_o \cdot \frac{\pi - \frac{\pi}{6}}{2\pi} = 9.32 \cdot \frac{5\pi}{6 \times 2\pi}$$

$$I_{T.A.} = \frac{9.32 \times 5}{12} = 3.883 \text{ A} \qquad \boxed{\therefore I_{T.A.} = 3.883 \text{ A}}$$

and r.m.s. current through thyristor is equal to :

$$I_{T.r} = \sqrt{I_o^2 \left(\frac{\pi - \frac{\pi}{6}}{2\pi} \right)} = I_o \sqrt{\frac{5\pi}{6 \times 2\pi}}$$

or
$$I_{T.r} = I_o \sqrt{\frac{5}{12}}$$

$$\boxed{I_{T.r} = 6.016 \text{ Amp}}$$

Average current through diode is euqal to :

$$I_{D.A.} = I_o \cdot \frac{\pi + \frac{\pi}{6}}{2\pi} = I_o \cdot \frac{7\pi}{12\pi} = 9.32 \times \frac{7}{12} = 5.436 \text{ Amp}$$

or
$$\boxed{I_{D.A.} = 5.436 \text{ Amp}}$$

r.m.s. current through diode is equal to :

$$I_{D.r} = I_o \sqrt{\frac{\pi + \frac{\pi}{6}}{2\pi}} = I_o \sqrt{\frac{7}{12}} = 9.32 \sqrt{\frac{7}{12}}$$

$$\boxed{I_{D.r} = 7.1182 \text{ Amp}}$$

(d) r.m.s. value of source current

$$I_{S.r} = I_o \sqrt{\frac{\pi - \alpha}{\pi}} = I_o \sqrt{\frac{\pi - \dfrac{\pi}{6}}{\pi}} = 9.32 \sqrt{\frac{5}{6}} = 8.508 \text{ Amp}$$

r.m.s value of load current $I_{or} = I_o = 9.32$ A

So, power delivered to load $= EI_o + I_{or}^2 R$

$$= 100 \times 9.32 + (9.33)^2 \times 10 = 1800.624 \text{ W}$$

$\therefore \ V_S I_S \cos \phi = 1800.624$ W

or $230 \times 8.508 \cos \phi = 1800.624$ W

or $\qquad \cos \phi = \dfrac{1800.624}{230 \times 8.508}$

$$\boxed{\text{p.f} = 0.9202 \text{ lag}}$$

(e) The circuit turn-off time is :

$$t_C = \frac{\pi - \alpha}{\omega} = \frac{\pi - \dfrac{\pi}{6}}{2\pi \times 50} \text{ sec} = \frac{5\pi}{6 \times 2\pi \times 50} \text{ sec} = \frac{5}{600} \text{ sec}$$

$$\boxed{t_C = 8.33 \text{ ms}}$$

Problem 4.19: Two $3 - \phi$ full converters are connected in antiparallel to form a $3 - \phi$ dual converter of the circulating-current type. The input to the dual converter is $3 - \phi$, 400 V, 50 Hz. If peak value of circulating current is to be limited to 20 A, find the value of inductance needed for the reactor for firing angles of (1) $\alpha_1 = 30°$, and (ii) $\alpha_1 = 60°$.

Solution: For a $3 - \phi$ full converter, the peak value of circulating current

occurs when $\omega t = \dfrac{5\pi}{6}$. This peak value of circulating current is given by:

$$\boxed{i_{CP} = \frac{\sqrt{3} V_{ml}}{\omega L} (1 - \sin \alpha_1)}$$

(1) at $\alpha_1 = 30°$

$i_{CP} = 20$ A, $V_{ml} = \sqrt{2} \times 400 = 565.68$ V, $\omega = 100 \pi$

So, $20 = \dfrac{\sqrt{3} \times \sqrt{2} \times 400}{\omega L} (1 - \sin 30^\circ)$

or $20 = \dfrac{979.79}{100\,\pi L} (1 - 0.5)$

or $L = \dfrac{979.79 \times 0.5}{100\,\pi \times 20}$

or $L = 0.07796$ H

or $\boxed{L = 77.96 \text{ mH}}$

(2) at $\alpha_1 = 60^\circ$

$i_{CP} = 20$ A, $V_{ml} = \sqrt{2} \times 400$ V, $\omega = 100\,\pi$

So, $20 = \dfrac{\sqrt{3} \times \sqrt{2} \times 400}{100\,\pi L} (1 - \sin 60^\circ)$

$L = \dfrac{\sqrt{3} \times \sqrt{2} \times 400}{100\,\pi \times 20} (1 - 0.866)$

$L = 0.02089$ H

or $\boxed{L = 20.89 \text{ mH}}$

Problem 4.20: The single phase half controlled a.c. to d.c. bridge converter of Fig. P. 4.18 supplies a 10 Ω resistor in series with a 100 V back e.m.f. load. The firing angle of the thyristor is set to 60°.
(a) Find the average current through the resistor.
(b) What will be the new average current through the resistor, if a very large inductor is connected in series with the load ?

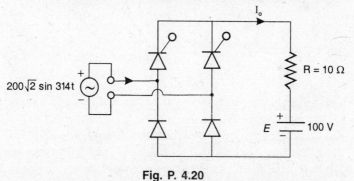

Fig. P. 4.20

Solution: For a 1 – φ half controlled bridge converter the average output voltage is given as :

$$V_o = \frac{V_m}{\pi}(1 + \cos\alpha) = \frac{200\sqrt{2}}{\pi}(1 + \cos 60^\circ)$$

$$\boxed{V_o = 135.04 \text{ V}}$$

(a) The average current through the resistor is given as :

$$I_o = \frac{V_o - E}{R} = \frac{135.04 - 100}{10}$$

or $\qquad \boxed{I_o = 3.5 \text{ Amp}}$

(b) If a very large inductor is connected in series with the load, current i_o rise gradually but its maximum value is remain as 3.5 Amp as previous one. This large inductance provide ripple free current.

Problem 4.21: A line commutated a.c. to d.c. converter is shown in Fig. P. 4.19. It operates from a 3 – φ, 50 Hz, 580 V (line to line) supply. It supplies a load current, I_0 of 3464 A. Assume I_0 to be ripple free and neglect source inpedance.

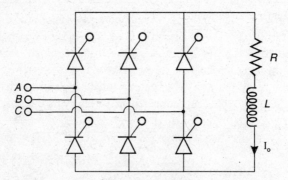

Fig. P. 4.21

(a) Calculate the delay angle α of the converter if its average output voltage is 648 V.
(b) Calculate the power delivered to the load R in kW.
(c) Sketch the wave form of the supply current i_A V_S time (m.sec).
(d) Calculate fundamental reactive power drown by converter from the supply in KVAR.

Solution: (a) Average output voltage is given as

$$V_o = \frac{3V_m}{\pi} \cos \alpha$$

$$648 = \frac{3}{\pi} \left(580 \times \sqrt{2}\right) \cos \alpha$$

or $\quad\quad \cos \alpha = 0.827$

or $\quad\quad \alpha = \cos^{-1}(0.827) = 34.16°$

or $\quad$ | delay angle $\alpha = 34.16°$ |

(b) Power delivered to the load R in kW is given as
$$P = V_o \times I_o = 648 \times 3464 = 2244.67$$

(c)

$$60° + 40° = 94° = \frac{90\pi}{180} = 0.52 \; \pi$$

For $\pi = 10$ ms, $\therefore 0.52 \; \pi = 5.2$ ms

(d) a.c. terminal power $P_{r.m.s.} = 1.05 \; P_{d.c.}$ (approx)
$$= 1.05 \times 2244.67 = 2356.9 \; KVA$$

$\therefore$ Power drawn by converter from the supply in KVAR or reactive power drawn is given as

$$\theta = \sqrt{(2356.9)^2 - (2244.67)^2}$$

| $\theta = 718.64 \; KVAR$ |

Problem 4.22: A 2 pulse phase controlled midpoint converter feeds on R.L. load. The load inductance is infinitely large to cause perfect smoothing. The load resistance is 20 Ω. The secondary voltage of the converter transformer is 230 V. Assuming the transformer and thyristors

to be ideal, determine the average values of load voltage and load current for firing angle of $\alpha = 30°$, $60°$. Determine the rating of thyristor.

Solution: For $\alpha = 30°$.

The average value of load voltage is given by

$$V_o = \frac{2\sqrt{2}}{\pi} \text{ V } \cos \alpha = 0.9 \times 230 \cos 30° = 179 \text{ V}$$

The average current $I_o = \frac{179}{20} = 8.95$ Amp

For $\alpha = 60°$

The average value of load voltage is

$$V_o = 0.9 \times 230 \times \cos 60° = 103.5 \text{ V}$$

The average current $I_o = \frac{103.5}{20} = 5.175$ Amp

	$\alpha = 30°$	$\alpha = 60°$
I_o	4.45	2.587
$I_{r.m.s.}$	6.29	3.65
$\therefore$ PIV of thyristor	325.22 V	325.22 V ($\sqrt{2}$ V)

Problem 4.23: A three pulse mid-point converter is supplied from a 400 V, $3 - \phi$ 50 Hz supply. It feeds a pure resistance of 2.5 Ω. At a firing angle of 120°, determine :
(a) The r.m.s. and average values of dc current.
(b) The r.m.s. and average value of thyristor current.
(c) The input power factor.

Solution: The average value of d.c. voltage

$$V_o = \frac{3\sqrt{2}}{2\pi} \text{ V } \{1 + \cos (\alpha + 30°)\}$$

or $$V_o = \frac{3\sqrt{2} \times 400}{2\pi} (\cos (120° + 30°) + 1)$$

$$\boxed{V_o = 20.9 \text{ V}}$$

The average value of d.c. current = 8.36 Amp.
r.m.s. value of voltage

$$V_{r.m.s.} = \sqrt{3} \cdot \sqrt{2}V \left[\frac{5}{24} - \frac{\alpha}{4\pi} + \frac{1}{8\pi} \sin\left(\frac{\pi}{3} + 2\alpha\right) \right]^{\frac{1}{2}}$$

$$\boxed{V_{r.m.s.} = 55.02 \text{ V}}$$

r.m.s. value of the current

$$I_{r.m.s.} = \frac{55.02}{2.5} = 22 \text{ A}$$

Average value of thyristor current

$$I_o = \frac{14.48}{3} = 4.82 \text{ Amp}$$

r.m.s. value of thyristor current = 12.71 Amp

Power delivered to load = $\sqrt{3} \times 12.71 \times 55.02 = 1211.2$ W

Apparent power = $\sqrt{3} \times 400 \times 12.71 = 85.05.5$ W

∴ Power factor = 0.137

Problem 4.24: A 3 pulse converter operates into an R-L load having an additional d.c. voltage of 250 V. The load resistance is 10 Ω and the inductance is large enough to provide ripple free current. The converter has a firing angle of 135°. Determine
(a) The d.c. load voltage.
(b) The average value of d.c. current.
(c) The thyristor PIV.
The converter transformer has a secondary voltage of 220 V at 50 Hz.

Solution: The average value of d.c. current of a converter is

$$I_o = \frac{V_o + E}{R}$$

(a) $\qquad V_o = \frac{3}{2\pi} \times 220\sqrt{2} \cos\alpha = \frac{3}{2\pi} \times 220\sqrt{2} \cdot \cos(135°)$

$$= -105.08 \text{ V}$$

(b) $\therefore$ $I_d = \dfrac{-105.08 + 142}{1.0}$

$$\boxed{I_d = 37 \text{ Amp}}$$

(c) Peak inverse voltage of thyristor = $220 \sqrt{2}$ V

$$\boxed{PIV = 311.12 \text{ V}}$$

Problem 4.25: A single phase bridge converter feeds an R-L load having a resistance of 5.5 Ω and an inductance of a very large value causing perfect smoothing. The converter is fed from a 400 V, 50 Hz single phase supply. For a firing angle of $\alpha = 75^\circ$ determine.
(a) The average value of output current.
(b) The r.m.s. value of output current.
(c) The average and r.m.s. thyristor currents.
(d) The power factor of the a.c. source.

Solution: The average value of d.c. voltage of the converter

$$V_o = 0.9 \text{ V} \cos \alpha = 0.9 \times 400 \times \cos 75^\circ$$

or $V_o = 93.18$ V

(a) So, the average value of d.c. current

$$I_o = \frac{V_o}{R} = \frac{93.18}{5.5} = 16.93 \text{ Amp}$$

(b) r.m.s. value of output current = 16.93 Amp

(c) The average value of thyristor current = $\dfrac{16.93}{2}$ = 8.46 Amp

The r.m.s. value of thyristor current = $\dfrac{16.93}{\sqrt{2}}$ = 11.97 A

(d) The power factor of a.c. source = 0.9 cos α
$$= 0.9 \times \cos 75^\circ = 0.233$$

Problem 4.26: A single phase bridge converter with a free wheeling diode feeds an R-L load. The load resistance in 10 Ω and inductance is very large providing ripple free load current. The converter is supplied by 220 V, 1 – ϕ supply at a frequency of 50 Hz. Determine the average value of load current, device current, power factor at a firing angle of 60°.

Solution: Average output voltage

$$V_o = \frac{2\sqrt{2}}{2\pi} \text{ V } (1 + \cos \alpha) = \frac{V_m}{\pi}(1 + \cos \alpha)$$

$$= \frac{\sqrt{2} \times 220}{\pi}(1 + \cos 60°) = \frac{297}{2}$$

$$\boxed{V_o = 148.55 \text{ V}}$$

The average value of load current
$$I_o = 14.855 \text{ A}$$
Average value of thyristor current

$$= I_o \left(\frac{\pi - \alpha}{2\pi}\right) = 14.855 \left(\frac{\pi - \dfrac{\pi}{3}}{2\pi}\right) = \frac{14.855}{3} = 4.95 \text{ Amp}$$

Average value of diode current

$$= I_o \left(\frac{\alpha}{\pi}\right) = 14.855 \frac{\pi}{3 \times \pi} = \frac{14.855}{3} = 4.95 \text{ Amp}$$

r.m.s. value of thyristor current = $4.95 \sqrt{3}$ = 8.57 Amp
r.m.s. value of diode current = 8.57 Amp

r.m.s. value of line current = $8.57 \sqrt{2}$ = 12.11 Amp
Power from the supply = 12.11 × 220 = 2664.4 W

$$\text{Power factor} = \frac{148.55 \times 14.855}{2664.2} = 0.828$$

Problem 4.27: A $3 - \phi$ half controlled converter supplies an *R-L* load with a ripple free current flowing in the load. The resistance in the load circuit is 2.5 Ω. The supply voltage to the converter is 120 V, $3 - \phi$ 50 Hz. Determine the average value of load current for a firing angle of 120°. Determine also the power delivered to load, the average and r.m.s. value of device currents. What is the line current of the supply ?

Solution: For a half controlled converter, the average output voltage is given by

$$V_o = \frac{3V_m}{\pi} \ (1 + \cos \alpha)$$

For $\alpha = 120^\circ$

$$V_o = \frac{3V_m}{\pi} \ (1 + \cos 120^\circ) = \frac{3\sqrt{2} \times 120}{\pi} \times 0.5 = 40.53 \text{ V}$$

Average value of load current $= \dfrac{40.53}{2.5} = 16.21 \text{ Amp}$

Power delivered to the load $= 657.05$ W

Average value of thyristor current $= 5.40$ Amp

r.m.s. value of thyristor current $= 9.36$ Amp

The line current $= I_o \sqrt{1 - \left(\dfrac{\alpha}{\pi}\right)} = \dfrac{I_0}{\sqrt{3}} = 9.36 \text{ Amp}$

Problem 4.28: Two $3 - \phi$ full converters are connected in antiparallel to form a $3 - \phi$ dual converter of the circulating current type. The input to the dual converter is $3 - \phi$, 400 V, 50 Hz. If the peak value of circulating current is to be limited to 15 A, find the value of inductance needed for the reactor for firing angle of (1) $\alpha_1 = 30^\circ$, and (2) $\alpha_1 = 45^\circ$.

Solution: For a $3 - \phi$ full converter, the peak value of circulating current occurs when $\omega t = \dfrac{5\pi}{6}$. The peak value of circulating current is given by :

$$\boxed{i_{CP} = \frac{\sqrt{3} \, V_{ml}}{\omega L} \ (1 - \sin \alpha_1)}$$

(1) For, $\alpha_1 = 30^\circ$

$$i_{CP} = 15 \text{ A}, \ V_{ml} = \sqrt{2} \times 400 = 565.88 \text{ V}, \ \omega = 100 \, \pi$$

So, $15 = \dfrac{\sqrt{3} \times \sqrt{2} \times 400}{100\pi L} \ (1 - \sin 30^\circ)$

$$L = \frac{979.78 \times 0.5}{100\pi \times 15}$$

$$L = 0.104 \text{ H}$$

or $\boxed{L = 104.5 \text{ mH}}$

(2) For $\alpha_1 = 45°$

$$i_{CP} = 15 \text{ A}, \ V_{ml} = \sqrt{2} \times 400 = 565.68 \text{ V}$$
$$\omega = 100 \ \pi$$

So, $$15 = \frac{\sqrt{3} \times \sqrt{2} \times 400}{100\pi L} \ (1 - \sin 45°)$$

or $$L = \frac{\sqrt{3} \times \sqrt{2} \times 400}{100\pi \times 15} \ (1 - 0.707)$$

$$L = \frac{979.78 \times 0.29}{100\pi \times 15}$$

$$L = 0.0608 \text{ H}$$

or $\boxed{L = 60.8 \text{ mH}}$

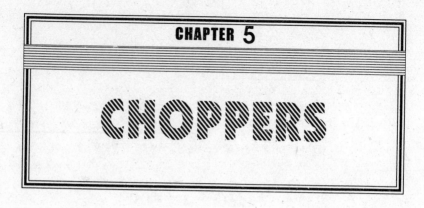

CHAPTER 5

CHOPPERS

A d.c. chopper converts fixed d.c. input voltage to a controllable d.c. output voltage.The chopper circuit require forced or load commutation to turn-off the thyristors. Classification of chopper circuits is dependent upon the type of commutation and also on the direction of power flow. Chopper are widely used in d.c. drives, battery-driven vehicles, subway cars, trolley, trucks etc.

A chopper may be through of as d.c equivalent of an a.c. tranformer since they behave in an identical manner. Chopper system offer smooth control, high efficiency, fast response and regeneration.

Principle of Chopper Operation

A chopper is high speed on/off semiconductor switch. It connects source to load and disconnects the load from source at a fast speed. It is of two types (1) step down chopper (2) step up chopper.

1. Step down chopper

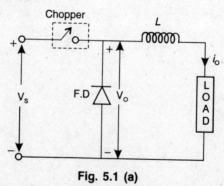

Fig. 5.1 (a)

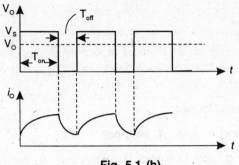

Fig. 5.1 (b)

$$V_o = \frac{T_{\text{on}}}{T_{\text{on}} + T_{\text{off}}} V_S$$

$$\boxed{V_o = \alpha\, V_S,}$$ where $\alpha = \dfrac{T_{\text{on}}}{T_{\text{on}} + T_{\text{off}}} = \dfrac{T_{\text{on}}}{T}$

α = duty cycle

The average value of output voltage V_o can be controlled through α (duty cycle). The various control strategies for varying duty cycle α are :

(a) Constant frequency system

In this method T_{on} time is varied but chopping frequency f is kept constant. It is also known as pulse width modulation scheme.

(b) Variable frequency system

In this method T_{on} or T_{off} is kept constant and the chopping frequency f is varied. This method of controlling α is also called frequency-modulation scheme.

2. Step-up choppers

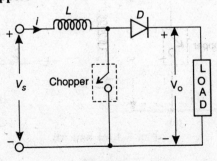

Fig. 5.2

$$V_o = \frac{T}{T - T_{on}} V_S \quad \text{or,} \quad V_o = \frac{1}{1 - \alpha} V_S$$

The method of commutation required in step-up choppers are generally force commutation. The principle of step-up chopper can be employed for the regenerative braking of d.c. motors.

Types of chopper circuits

(a) First-quadrant or type A, chopper

The power flow in type A chopper is always from source to load. This chopper is also called step-down chopper as average output voltage V_o is always less than the input d.c voltage V_S.

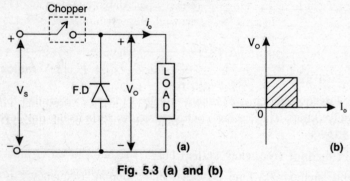

Fig. 5.3 (a) and (b)

(b) Second-quadrant or type B, chopper

The power flow in type B chopper is always from load to source. As load voltage V_o is more than source voltage V_S, type B chopper is also called step-up chopper.

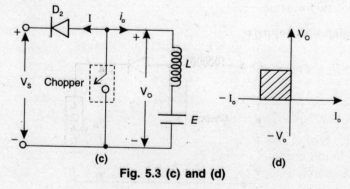

Fig. 5.3 (c) and (d)

(c) Two quadrant type A chopper or type C chopper

In this type of circuit power flow may be from source to load (first quadrant) or from load to source (second quadrant) choppers CH1 and CH2 should not be on simultaneously as this would lead to a direct short circuit on the supply line. This type of chopper is generally used for motoring and regenerative braking of d.c. motors.

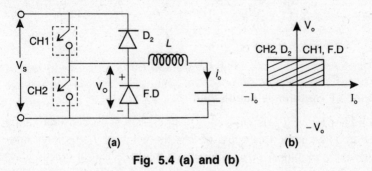

(a)   (b)

Fig. 5.4 (a) and (b)

(d) Two quadrant type B chopper, or type D chopper

The direction of load current in this type of chopper is always positive because chopper and diodes can conduct current only in the direction of arrows.

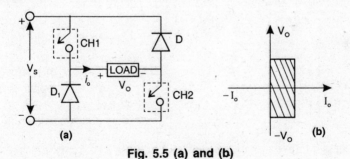

(a)   (b)

Fig. 5.5 (a) and (b)

$V_o = +ve$ when $T_{on} > T_{off}$

$V_o = -ve$ when $T_{on} < T_{off}$

(e) Four quadrant chopper or type E chopper

In this type of chopper power is fed back from load to source in second and fourth quadrant. In first and Third quadrant power is flow always from source to load.

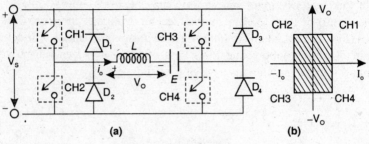

Fig. 5.6 (a) and (b)

Limit of continuous conduction

In a chopper, if T_{on} is reduced, T_{off} in creases for a constant chopping period T. At some low value of T_{on}, the value of T_{off} is large and the current i may fall to zero. Since the current reached in type A chopper can not reverse, it stays at zero. The limit of continuous conduction is reached when I_{mn} goes to zero. The value of duty cycle α at the limit of continuous conduction is obtained by equating I_{mn} to zero, i.e.

$$I_{mn} = \frac{V_S}{R}\left[\frac{e^{\frac{T_{on}}{T_a}}-1}{e^{\frac{T}{Ta}}-1}\right] - \frac{E}{R} = 0$$

or

$$\left[\frac{e^{\frac{T_{on}}{T_a}}-1}{e^{\frac{T}{Ta}}-1}\right] = \frac{E}{V_S} = k$$

or

$$e^{\frac{T_{on}}{T_a}} = 1 + k\left(e^{\frac{T}{Ta}}-1\right)$$

or

$$\alpha' = \frac{T_{on}}{T} = \frac{T_a}{T}\ln\left[1 + k\left(e^{\frac{T}{Ta}}-1\right)\right]$$

(i)   The straight line AC and ACD is not possible as $\dfrac{T_a}{T}$ can never be less than zero.

(ii) The straight line AC and curve $AA'C$ represent discontinuous current mode for $0 < \dfrac{T_a}{T} < 1$.

(iii) Curve $AA'C$ and ABC represents continuous current mode for $\dfrac{T_a}{T} > 1$.

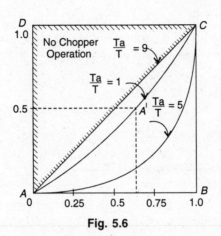

Fig. 5.6

Type of Commutation in Thyristor Chopper Circuit

(a) **Forced commutation :** In forced commutation, external element L and C does not carry load current continuously are used to turn-off a conducting thyristor. Forced commutation can be acheived in the following two ways.

(i) *Voltage commutation :* In this scheme, a conducting thyristor is commutated by the application of a pulse of large reverse voltage. The sudden application of reverse voltage across the conducting thyristor reduces the anode current to zero rapidly.

(ii) *Current commutation :* In this scheme, an external pulse of current greater than the load current is passed in the reversed direction through the conducting SCR. When the current pulse attains a value equal to the load current, net pulse current through thyristor becomes zero and the device is turned off. Due to the reverse diode in current commutation, the commutation time in current commutation is more as compared to that in voltage commutation.

In both voltage and current commutation schemes, commutation is initiated by gating an auxiliary SCR.

(b) **Load commutation :** In load commutation, a conducting thyristor is turned off when load current flowing through a thyristor either.

(i) Becomes zero due to the nature of load circuit parameters.

(ii) Is transferred to another device from the conducting thyristor.

Points to be Noted

(1) For a chopper, V_S is the source voltage, R is the load resistance and α is the duty cycle. r.m.s. and average values of thyristor currents for this chopper are $\sqrt{\alpha}\ \dfrac{V_S}{R}$ and $\alpha\ \dfrac{V_S}{R}$ respectively.

(2) In d.c. chopper, per unit ripple is maximum when duty cycle α is 0.5.

$$\text{Per unit ripple current} = \frac{I_{mx} - I_{nx}}{\dfrac{V_S}{R}} = \frac{\left(1 - e^{-\frac{\alpha T}{Ta}}\right)\left(1 - e^{-\frac{(1-\alpha)T}{Ta}}\right)}{1 - e^{-\frac{T}{Ta}}}$$

$$T_a \propto L \quad [\because T_a = \frac{L}{R}]$$

T_a increase then p.u ripple current decreases.

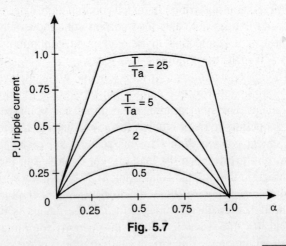

Fig. 5.7

(3) The amplitude of harmonic voltage is equal to $\dfrac{2V_S}{n\pi} \sin n\pi\ \alpha$

Ripple factor $R \cdot F = \dfrac{V_S}{V_o} = \dfrac{V_S \sqrt{\alpha - \alpha^2}}{V_S \cdot \alpha} = \sqrt{\dfrac{1 - \alpha}{\alpha}} = \sqrt{\dfrac{1}{\alpha} - 1}$

$\therefore$

$$\boxed{R \cdot F = \sqrt{\dfrac{1}{\alpha} - 1}}$$

(4) When a series LC circuit is connected to a d.c. supply of V volts through a thyristor, then the peak current through thyristor is equal to

$V_S \sqrt{\dfrac{C}{L}}$.

PROBLEMS

Problem 5.1: A d.c. battery is charged from a constant d.c. source of 220 V through a chopper. The d.c. battery is to be charged from its internal emf of 90 V to 122 V. The battery has internal resistance of 1 Ω. For a constant charging current of 10 A, compute the range of duty cycle.

Solution: Output voltage of a chopper is equal to :

$$\boxed{V_o = \alpha \, V_S}$$

where α = duty cycle

For a battery charging: Average output current is equal to :

$$I_o = \frac{V_o - E}{R}$$

Given, $I_o = 10$ A, $R = 1$ Ω, $E = 90$ V to 122 V
$V_S = 220$ V and $V_o = \alpha \, V_S$
So, For $E = 90$ V

$$10 \text{ A} = \frac{\alpha V_S - 90}{1}$$

or $\quad 10 + 90 = \alpha \times 220$

or $\quad \alpha = \dfrac{100}{220}$

or $\quad \boxed{\alpha = 0.4545}$

For $E = 122$ V

$$10 \text{ A} = \frac{\alpha' V_S - 122}{1}$$

or $10 \text{ A} + 122 = \alpha' \times 220$

or $\boxed{\alpha' = \dfrac{132}{220} = 0.6}$

So, range of duty cycle is equal to 0.4545 to 0.6.

Problem 5.2: For type A chopper, express the following variables in terms of V_S, R, I_o and duty cycle α in case load inductance causes the load current I_o to remain constant at a value $I_o = \dfrac{V_o}{R}$. Here V_S is the source voltage.

(a)   Average output voltage and current.

(b)   Output current at the instant of commutation.

(c)   Average and r.m.s. value of free wheeling diode current.

(d)   r.m.s. value of the output voltage.

(e)   Average and r.m.s. value of thyristor current. Sketch the time variations of gate single, output voltage V_o, output current i_o, thyristor current i_T and free wheeling diode current i_{fd}.

Solution:

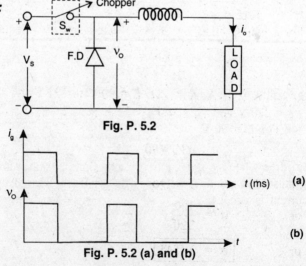

Fig. P. 5.2

Fig. P. 5.2 (a) and (b)

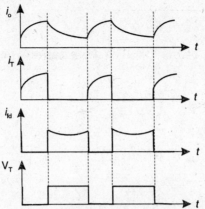

Fig. P. 5.2 (c - f) Wave form of i_g, V_o, I_o, i_T, i_{fD}, V_T, V_s, time t

(a) Average output voltage :

$$V_o = \frac{T_{on}}{T_{on} + T_{off}} V_S = \frac{T_{on}}{T} V_S$$

$$\boxed{V_o = \alpha \, V_S}$$

Average output current :

$$I_o = \frac{V_o}{R} = \frac{\alpha \, V_S}{R}$$

$\therefore$
$$\boxed{I_o = \frac{\alpha \, V_S}{R}}$$

(b) The output current is commutated by the thyristor at the instant $t = T_{on}$. Therefore, output current at the instant of commutation is $= \dfrac{V_o}{R}$

$= \dfrac{V_o}{R}$.

(c) Free wheeling diode current come into existance at the end of gating pulse and continue till next gating pulse is started. So, average F.D current is equal to :

$$I_{fd} = \frac{T_{off}}{T_{on} + T_{off}} I_o$$

$$\boxed{I_{fd} = 1 - \alpha \, I_o}$$

r.m.s. value of free wheeling diode current is equal to :

$$I_{fd,r} = \left(\frac{T_{\text{off}}}{T_{\text{on}} + T_{\text{off}}} I_o^2 \right)^{\frac{1}{2}}$$

or $\qquad I_{fd,r} = \left((1 - \alpha) \, I_o^2 \right)^{\frac{1}{2}}$

or $\qquad \boxed{I_{fd,r} = \sqrt{1 - \alpha} \, I_o}$

(d) r.m.s. value of output voltage is given as :

$$V_{or} = \left(\frac{T_{\text{on}}}{T_{\text{on}} + T_{\text{off}}} V_s^2 \right)^{\frac{1}{2}}$$

$$V_{or} = \left(\alpha V_S^2 \right)^{\frac{1}{2}}$$

or $\qquad \boxed{V_{or} = \sqrt{\alpha} \, V_S}$

(e) Average thyristor current is equal to :

$$i_{TAV} = \frac{T_{\text{on}}}{T_{\text{on}} + T_{\text{off}}} I_o$$

$$\boxed{i_{TAV} = \alpha \, I_o}$$

r.m.s. thyristor current is equal to :

$$i_{Tr} = \left(\alpha \, I_o^2 \right)^{\frac{1}{2}}$$

or $\qquad \boxed{i_{Tr} = \sqrt{\alpha} \, I_o}$

Problem 5.3: Draw the power circuit diagram for a type *A* chopper. Show load voltage wave form for (i) $\alpha = 0.3$, and (ii) $\alpha = 0.8$. For both these duty cycles, calculate :

(a) The average and r.m.s. value of output voltage in terms of source voltage V_S.

(b) The output power in case of resistive load R.

(c) The ripple factors.

Solution:

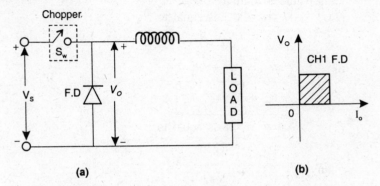

Fig. P. 5.3 (a) and (b). Circuit diagram of class A chopper

Load voltage waveform for $\alpha = 0.3$ and $\alpha = 0.8$ is given as :

(1) at $\alpha = 0.3$

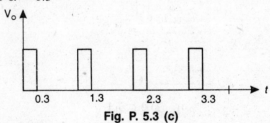

Fig. P. 5.3 (c)

(2) at $\alpha = 0.8$

Fig. P. 5.3 (d)

(a) Average value of output voltage at

(1) $\alpha = 0.3$ is equal to :

$$V_o = \alpha V_S = 0.3 \, V_S$$

∴ $$\boxed{V_o = 0.3 \ V_S}$$

(2) and at $\alpha = 0.8$ is equal to :

$$\boxed{V_o = 0.8 \ V_S}$$

r.m.s. value of output voltage at

(1) $\alpha = 0.3$ is equal to :

$$V_{or} = \sqrt{\alpha} \ V_S$$

or $$V_{or} = \sqrt{0.3} \ V_S$$

or $$\boxed{V_{or} = 0.5477 \ V_S}$$

(2)   at $\alpha = 0.8$ is equal to :

$$V_{or} = \sqrt{0.8} \ V_S$$

$$\boxed{V_{or} = 0.8944 \ V_S}$$

(b) Output power in case of resistive load R at

(1) at $\alpha = 0.3$ is equal to :

$$\frac{V_{or}^2}{R} = \frac{\left(\sqrt{\alpha} \ V_S\right)^2}{R} = \frac{\alpha V_S^2}{R}$$

$$P_o = \frac{0.3 V_S^2}{R}$$

(2) at $\alpha = 0.8$ is equal to :

$$P_o = \frac{\alpha V_S^2}{R} = \frac{0.8 V_S^2}{R}$$

$$\boxed{P_o = \frac{0.8 V_S^2}{R}}$$

(c) Ripple factors at

(1) $\alpha = 0.3$ is equal to :

$$R \cdot F = \sqrt{\frac{1-\alpha}{\alpha}} = \sqrt{\frac{1}{\alpha} - 1} = \sqrt{\frac{1}{0.3} - 1} = \sqrt{3.333 - 1}$$

$$= \sqrt{2.333} = 1.5275$$

$$\therefore \qquad \boxed{R \cdot F = 1.5275}$$

(2) at $\alpha = 0.8$ is equal to :

$$R \cdot F = \sqrt{\frac{1}{\alpha} - 1} = \sqrt{\frac{1}{0.8} - 1} = \sqrt{1.25 - 1} = \sqrt{0.25}$$

or $\qquad \boxed{R \cdot F = 0.5}$

Problem 5.4: A chopper has the following data $T = 1000$ µs, $R = 2$ Ω, $L = 5$ mH.
Find the duty cycle α so that per unit value of minimum load current does not fall below.

(i)   0.1, and (ii) 0.3 of $\dfrac{V_S}{R}$.

Solution: $\qquad T_a = \dfrac{L}{R} = \dfrac{5 \times 10^{-3}}{2} = 2.5 \times 10^{-3}$

$$\therefore \qquad \boxed{T_a = 2.5 \times 10^{-3}}$$

Duty cyle α so that P.u value of minimum load current does not fall below

(1) 0.1 of $\dfrac{V_S}{R}$:

$$\alpha = \frac{T_a}{T} \ln \left[1 + m \left(e^{\left(\frac{T}{T_a}\right)} - 1 \right) \right]$$

$$= \frac{2.5 \times 10^{-3}}{10^{-3}} \ln \left[1 + 0.1 \left(e^{\frac{10^{-3}}{2.5 \times 10^{-3}}} - 1 \right) \right]$$

$$= 2.5 \ln [1 + 0.1 \times (1.4918 - 1)]$$
$$= 2.5 \ln [1 + 0.0491] = 2.5 \times 0.048$$

$$\boxed{\alpha = 0.12}$$

(2) 0.3 of $\dfrac{V_S}{R}$:

$$\alpha = \frac{T_a}{T} \ln \left[1 + m \left(e^{\frac{T}{Ta}} - 1 \right) \right]$$

$$= \frac{2.5 \times 10^{-3}}{10^{-3}} \ln \left[1 + 0.3 \left(e^{\frac{10^{-3}}{2.5 \times 10^{-3}}} - 1 \right) \right]$$

$$= \frac{2.5 \times 10^{-3}}{10^{-3}} \ln [1 + 0.3 \times (1.4918 - 1)]$$

$$= 2.5 \ln [1 + 0.3 \times 0.4918]$$

$$= 2.5 \ln [1 + 0.14754] = 2.5 \times 0.1376$$

$$\boxed{\alpha = 0.344}$$

Problem 5.5: (a) A chopper fed from a 220 V d.c. source, is working at a frequency of 50 Hz and is connected to an R-L load of $R = 5\,\Omega$ and $L = 40$ mH. Determine the value of duty cycle at which the minimum load current will be (i) 5 A, (ii) 10 A, (iii) 20 A, (iv) 30 A. (b) For the values of α obtained in (a), calculate the corresponding values of maximum current and the ripple factors.

Solution:

(a)   Frequency = 50 Hz

$$\therefore \text{period } T = \frac{1}{f} = \frac{1}{50} = 20 \text{ m sec}$$

$$\text{time constant } T_a = \frac{L}{R} = \frac{40}{5} \text{ m sec}$$

or $T_a = 8$ m sec

(i) At minimum load current of 5 A

$$m = \frac{5A}{\dfrac{V_S}{R}} = \frac{5}{\dfrac{220}{5}} = \frac{5}{44}$$

$$\therefore \qquad \boxed{m = \frac{5}{44}}$$

duty cycle $\alpha' = \dfrac{T_a}{T} \ln\left(1 + m\left(e^{\frac{T}{T_a}} - 1\right)\right)$

$$\therefore \qquad \alpha' = \frac{8}{20} \ln\left(1 + \frac{5}{44}\left(e^{\frac{20}{8}} - 1\right)\right) = \frac{8}{20} \ln\left(1 + \frac{5}{44} \times 11.18\right)$$

$$= \frac{1}{2.5} \ln(1 + 1.27) = \frac{0.820}{2.5} = 0.328$$

$$\boxed{\alpha' = 0.328}$$

(ii) At minimum load current of 10 A

$$m = \frac{10}{44}$$

$\therefore$ duty cycle $\alpha' = \dfrac{T_a}{T} \ln\left(1 + m\left(e^{\frac{T}{T_a}} - 1\right)\right)$

$$= \frac{8}{20} \ln\left(1 + \frac{10}{44} \times 11.18\right) = \frac{1}{2.5} \ln(3.54)$$

$$= \frac{1.2643}{2.5} = 0.505$$

$$\therefore \qquad \boxed{\alpha' = 0.505}$$

(iii) At minimum load current of 20 A

$$m = \frac{20}{44}$$

$$\therefore \qquad \alpha' = \frac{1}{2.5} \ln\left(1 + \frac{20}{44} \times 11.18\right) = \frac{1}{2.5} \ln(6.08)$$

$$= \frac{1.8}{2.5} = 0.722$$

$$\therefore \qquad \boxed{\alpha' = 0.722}$$

(iv) At minimum load current of 30 A

$$m = \frac{30}{44}$$

$$\therefore \qquad \alpha' = \frac{1}{2.5} \ln \left(1 + \frac{30}{44} \times 11.18\right) = \frac{1}{2.5} \ln (8.622)$$

$$= \frac{2.1544}{2.5} = 0.8617$$

$$\therefore \qquad \boxed{\alpha' = 0.8617}$$

(b)   (i) For $\alpha = 0.328 = \dfrac{T_{\text{on}}}{T}$

$$T_{\text{on}} = 0.328 \times 20 \text{ m sec}$$
$$T_a = 8 \text{ m sec}$$

$\therefore$ Maximum current I_{mx} is given as :

$$I_{mx} = \frac{V_S}{R} \left[\frac{1 - e^{\frac{T_{\text{on}}}{T_a}}}{1 - e^{-\frac{T}{T_a}}}\right]$$

or $\qquad I_{mx} = \dfrac{220}{5} \left[\dfrac{1 - e^{-\frac{0.328 \times 20}{8}}}{1 - e^{-\frac{20}{8}}}\right] = 44 \left[\dfrac{1 - e^{-0.82}}{1 - e^{-2.5}}\right]$

$$= 44 \frac{[1 - 0.440]}{[1 - 0.082]} = 44 \times \frac{0.56}{0.9179} = 26.84 \text{ Amp}$$

$$\therefore \qquad \boxed{I_{mx} = 26.84 \text{ Amp}}$$

ripple factor at $\alpha = 0.328$ is :

$$R \cdot F = \sqrt{\frac{1-\alpha}{\alpha}} = \sqrt{\frac{1-0.328}{0.328}}$$

$$\boxed{R \cdot F = 1.4313}$$

(ii) at $\alpha = 0.50582$

$T_{on} = 0.50582 \times 20$ m sec

$\therefore$ $$I_{mx} = 44 \frac{\left[1 - e^{-0.50582 \times 2.5}\right]}{\left[1 - e^{-2.5}\right]} = 44 \frac{[0.7176]}{[0.9179]}$$

$$\boxed{I_{mx} = 34.4 \text{ Amp}}$$

$$R \cdot F = \sqrt{\frac{1}{\alpha} - 1} = \sqrt{\frac{1}{0.50582} - 1}$$

$$\boxed{R \cdot F = 0.9884}$$

(iii) at $\alpha = 0.7222$

$T_{on} = 0.7222 \times 20$ m sec

$\therefore$ $$I_{mx} = 44 \frac{\left[1 - e^{-0.7222 \times 2.5}\right]}{\left[1 - e^{-2.5}\right]} = 44 \frac{[0.8356]}{[0.9179]}$$

$$\boxed{I_{mx} = 40.05 \text{ Amp}}$$

$$R \cdot F = \sqrt{\frac{1 - 0.7222}{0.7222}}$$

$$\boxed{R \cdot F = 0.62}$$

(iv) at $\alpha = 0.86184$

$T_{on} = 0.86184 \times 20$ m sec

$\therefore$ $$I_{mx} = 44 \frac{\left[1 - e^{-0.86184 \times 2.5}\right]}{\left[1 - e^{-2.5}\right]} \quad \text{or} \quad I_{mx} = 44 \times \frac{0.8840}{0.9179}$$

$$\boxed{I_{mx} \ = \ 42.377 \ Amp}$$

$$R{\cdot}F \ = \ \sqrt{\frac{1-0.86184}{0.86184}}$$

$$\boxed{R{\cdot}F \ = \ 0.4004}$$

Problem 5.6: (a) A d.c. chopper feeds power to an RLE load with $R = 2 \ \Omega$, $L = 10$ mH and $E = 6$ V. If this chopper is operating at a chopping frequency of 1 kHz and with duty cycle of 10 % from a 220 V d.c. source, compute the maximum and minimum currents taken by the load.

(b)   A d.c. chopper is used to control the speed of a separately excited d.c. motor. The d.c. supply voltage is 220 V, armature resistance r_a

$= 0.2 \ \Omega$ and motor constant $k_a \ \phi = 0.8 \ \dfrac{V}{\text{r.p.m.}}$.

The motor drives a constant torque load requiring an average armature current of 25 A. Determine:
(i) The range of speed control, (ii) Range of duty cycle. Assume the motor current to be continuous.

Solution: (a) Given : $R = 2 \ \Omega$, $L = 10$ mH, $E = 6$ V, $V_S = 220$ V

$$\alpha = 0.1, \ T = 1 \ m \ sec, \ T_A = \frac{L}{R} = \frac{10}{2} m \ sec = 5 \ m \ sec$$

$$\therefore \qquad T_{on} = \alpha T = 0.1 \ m \ sec$$

Maximum current taken by the load

$$I_{mx} \ = \ \frac{V_S}{R}\left[\frac{1-e^{-\frac{T_{on}}{T_A}}}{1-e^{-\frac{T}{T_A}}}\right] - \frac{E}{R}$$

$$\therefore \qquad I_{mx} \ = \ \frac{220}{2}\left[\frac{1-e^{-\frac{0.1}{5}}}{1-e^{-\frac{1}{5}}}\right] - \frac{6}{2} = 110\left[\frac{1-0.980}{1-0.818}\right] - \frac{6}{2}$$

$$I_{mx} = 110 \times \frac{0.02}{0.182} - 3 = 12.015 - 3$$

$$\boxed{I_{mx} = 9.015 \text{ Amp}}$$

Minimum current taken by the load

$$I_{mn} = \frac{V_S}{R} \left[\frac{e^{\frac{T_{on}}{T_a}} - 1}{e^{\frac{T}{T_a}} - 1} \right] - \frac{E}{R}$$

$$= \frac{220}{2} \left[\frac{e^{\frac{0.1}{5}} - 1}{e^{\frac{1}{5}} - 1} \right] - \frac{6}{2} = 110 \left[\frac{1.02020 - 1}{1.221 - 1} \right] - 3$$

$$= 110 \times 0.0914 - 3 = 10.0549 - 3$$

$$\boxed{I_{mn} = 7.054 \text{ Amp}}$$

(b) For a motor, $V_t = V_o = E_a + I_a r_a$.

The minimum possible speed of dc motor is zero. This gives motor counter emf $E_a = 0$.

$$\therefore \qquad \alpha V_S = V_o = 0 + I_a r_a$$

or $\qquad \alpha \times 220 = 25 \times 0.2$

or $\qquad \alpha = \frac{5}{220} = \frac{1}{44}$

Maximum possible value of duty cycle 1

$$\therefore \qquad \alpha V_S = E_a + I_a r_a$$

$$1 \times 220 = 0.8 \times N + 25 \times 0.2$$

$$220 - 5 = 0.8 N$$

or $\qquad N = \frac{215}{0.8} = 268.75$ r.p.m.

So range of duty cycle is $\frac{1}{44} < \alpha < r$ and range of speed : $0 < N < 268.75$ r.p.m.

Problem 5.7: A step-up chopper has output voltage of two to four times the input voltage. For a chopping frequency of 2000 Hz, determine the range of off-periods for the date signal.

Solution: For step up chopper

$$V_o = \frac{V_S}{1-\alpha}, \qquad \text{Given } f = 2000 \text{ Hz}, \therefore T = \frac{1}{f} = 5 \times 10^{-4} \text{ sec}$$

When, $V_o = 2 V_S$

Then, $2 V_S = \dfrac{V_S}{1-\alpha}$

or $1 - \alpha = \dfrac{1}{2} = 0.5$

or $\alpha = 1 - 0.5 = 0.5$

$\therefore \qquad \boxed{\alpha = 0.5}$

$\because \qquad \alpha = \dfrac{T_{on}}{T_{off} + T_{on}} = \dfrac{T_{on}}{T} = \dfrac{T_{on}}{5 \times 10^{-4} \text{ sec}}$

$\therefore \qquad T_{on} = \alpha \times 5 \times 10^{-4} \text{ sec} = 0.5 \times 5 \times 10^{-4} \text{ sec}$

$$= 2.5 \times 10^{-4} \text{ sec} = 250 \times 10^{-6} \text{ sec}$$

$\boxed{T_{on} = 250 \ \mu s} \qquad\qquad \therefore T_{off} = (500 - 250) \ \mu s$

$\therefore \qquad \boxed{T_{off} = 250 \ \mu s}$

When, $V_o = 4 V_S$

Then, $4 V_S = \dfrac{V_S}{1-\alpha}$

or $1 - \alpha = 0.25$

$\therefore \qquad \alpha = 1 - 0.25$

$\boxed{\alpha = 0.75}$

$$T_{on} = \alpha T = 0.75 \times 5 \times 10^{-4} \text{ sec} = 3.75 \times 10^{-4} \text{ sec}$$

$\boxed{T_{on} = 375 \ \mu \text{ sec}}$

$\therefore \qquad T_{off} = 500 \ \mu s - 375 \ \mu \text{ sec}$

$\therefore \qquad \boxed{T_{off} = 125 \ \mu s}$

Hence, range of off periods for the gate signal is

$$250 \ \mu s \text{ to } 125 \ \mu s$$

Problem 5.8: For type A chopper feeding an RLE load, show that the maximum value of average current rating for the free wheeling diode, in case load current remains constant, is given by :

$$\left[\frac{V_S}{4R}\left(1-\frac{E}{V_S}\right)^2\right]$$

Solution: For type A chopper feeding an RLE load, average current rating for free wheeling diode is given as :

$$i_{fd} = (1-\alpha)\, I_o$$

$$i_{fd} = (1-\alpha)\left(\frac{\alpha V_S - E}{R}\right) \qquad ...(1)$$

$$i_{fd} = (\alpha - \alpha^2)\,\frac{V_S}{R} - (1-\alpha)\,\frac{E}{R}$$

$$\frac{di_{fd}}{d\alpha} = (1-2\alpha)\,\frac{V_S}{R} + \frac{E}{R} = 0$$

or $\qquad E = (2\alpha - 1)\, V_S$

or $\qquad \dfrac{E}{V_S} + 1 = 2\,\alpha$

or $\qquad \boxed{\alpha = \dfrac{E + V_S}{2V_S}}$

Putting the value of α in eq. (1) we get

$$(i_{fd})_{max} = \left(1 - \frac{E+V_S}{2V_S}\right)\left(\frac{\dfrac{E+V_S}{2V_S}\cdot V_S - E}{R}\right)$$

$$= \frac{V_S - E}{2V_S}\left(\frac{V_S - E}{2R}\right) = \frac{(V_S - E)^2}{4V_S R} = \frac{V_S^2\left(1 - \dfrac{E}{V_S}\right)^2}{4V_S R}$$

$$\boxed{(i_{fd})_{max} = \frac{V_S\left(1 - \dfrac{E}{V_S}\right)^2}{4R}}$$

Problem 5.9: A battery with its terminal voltage of 200 V is supplied with power from type A chopper circuit. The output voltage of the chopper consists of rectangular pulses of 2 ms duration in an overall cycle time of 5 ms. Internal resistance of the battery is negligible calculate :
(a) Ripple factor.
(b) Average and r.m.s. value of output voltage.
(c) r.m.s. value of the fundamental component of output voltage.
(d) a.c. ripple voltage.

Solution: (a) duty cycle $\alpha = \dfrac{2\ ms}{5\ ms} = \dfrac{2}{5}$

$$\therefore \qquad R \cdot F = \sqrt{\frac{1-\alpha}{\alpha}} = \sqrt{\frac{1-\dfrac{2}{5}}{\dfrac{2}{5}}} = \sqrt{\frac{3}{2}} = 1.2247$$

$$\therefore \qquad \boxed{R \cdot F = 1.2247}$$

(b) average output voltage

$$V_o = \alpha\, V_S = \frac{2}{5} \times 200$$

$$\boxed{V_o = 80\ V}$$

r.m.s. output voltage

$$V_{or} = \alpha\, V_S = \sqrt{0.4} \times 200 = 0.632 \times 200$$

$$\boxed{V_{or} = 126.49\ V}$$

(c) r.m.s. value of nth harmonic voltage is

$$V_n = \frac{2 V_S}{\sqrt{2}\ n\pi} \sin (n\pi\alpha)$$

For, first harmonic voltage is

$$V_1 = \frac{2 V_S}{\sqrt{2}\ \pi} \sin (\pi\alpha) = \frac{2 \times 200}{\sqrt{2}\ \pi} \sin \left(180 \times \frac{2}{5}\right)$$

$$\boxed{V_1 = 85.625 \text{ Volt}}$$

(d) a.c. ripple voltage

$$V_r = \sqrt{V_{or}^2 - V_o^2} = \sqrt{(126.49)^2 - (80)^2}$$

$$= \sqrt{9592.1316}$$

$$\boxed{V_r = 97.93 \text{ Volt}}$$

Problem 5.10: A type A chopper feeds power to RLE load with $R = 1.5\ \Omega$, $L = 6$ mH and $E = 44$ V. Other data for their chopper is as under :
Source voltage = 220 V d.c., chopping frequency = 1 kHz output voltage pulse duration = 400 μ sec. Find
(a) Find whether load current is continuous or not.
(b) Calculate the value of average output current.
(c) Compute the maximum and minimum values of steady state output current.
(d) Sketch the time variations of gate signal i_g. load voltage V_o, load current i_o, thyristor current i_T, free wheeling diode current i_{fd} and voltage across thyristor V_T.
(e) Find r.m.s. values of the first, second and third harmonics of the load current.
(f) Compute the average value of supply current.
(g) Compute input power, the power absorbed by the load counter e.m.f. and the power loss in the resistor.
(h) Computer r.m.s. value of load current using the results of (b) and (g).
(i) Using results of (e), find the r.m.s. value of load current. Compare the result with that obtained in part (h).

Solution: (a) $T_a = \dfrac{L}{R} = \dfrac{6\,\text{mH}}{1.5} = 4$ m sec

$$T = \frac{1}{1\,\text{kHz}} = 1 \text{ msec}$$

$$\frac{T_a}{T} = \frac{4}{1} = 4$$

and
$$\frac{T_a}{T} = \frac{1}{4} = 0.25$$

$$m = \frac{E}{V_S} = \frac{44}{220} = 0.2$$

$$\alpha = \frac{0.4}{1} = 0.4 \qquad\qquad \therefore T_{on} = 0.4$$

$$\therefore \quad \frac{T_{on}}{T_a} = \frac{0.4}{4} = 0.1$$

$$\therefore \quad \alpha' = \frac{T_a}{T} \ \ln\left(1 + m\left(e^{\frac{T}{T_a}} - 1\right)\right)$$

$$= 4 \ \ln\left(1 + 0.2\left(e^{0.25} - 1\right)\right) = 4 \ \ln\left(1 + 0.056\right)$$

$$\boxed{\alpha' = 0.221}$$

Since α (0.4) > α' (0.221), so the load current is continuous.
(b) Average output current is

$$I_o = \frac{\alpha V_S - E}{R} = \frac{0.4 \times 220 - 44}{1.5}$$

$$\boxed{I_o = 29.33 \ \text{Amp}}$$

(c) Maximum values of steady-state output current is

$$I_{mx} = \frac{V_S}{R}\left[\frac{1 - e^{\frac{T_{on}}{T_a}}}{1 - e^{-\frac{T}{T_a}}}\right] - \frac{E}{R} = \frac{220}{1.5}\left[\frac{1 - e^{-0.1}}{1 - e^{-0.25}}\right] - \frac{44}{1.5}$$

$$= \frac{220}{1.5}\frac{[0.0951]}{[0.2211]} - \frac{44}{1.5} = 63.05 - 29.33$$

$$\boxed{I_{mx} = 33.72 \ \text{Amp}}$$

Minimum values of steady state output current is

$$I_{mn} = \frac{220}{1.5}\left[\frac{e^{0.1} - 1}{e^{0.25} - 1}\right] - \frac{44}{1.5} = \frac{220}{1.5}\frac{(0.10517)}{(0.2840)} - \frac{44}{1.5}$$

$$= 54.30 - 29.33$$

$$\boxed{I_{mn} = 24.974 \text{ Amp}}$$

(d)

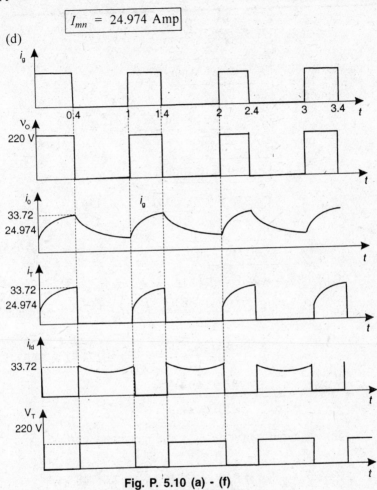

Fig. P. 5.10 (a) - (f)

(e) r.m.s. value of first harmonics voltage is

$$V_1 = \frac{2V_S}{\sqrt{2}\,\pi} \sin(\pi \times 0.4) = \frac{2 \times 220}{\sqrt{2}\,\pi} \sin 72°$$

$$\boxed{V_1 = 94.18 \text{ V}}$$

Chopping frequency = 1 kHz

$$\therefore \quad \omega = 2\pi f = 6.283 \times 10^3$$

$$\therefore \quad Z_1 = \sqrt{R^2 + (\omega L)^2} = \sqrt{(15)^2 + (6.283 \times 10^3 \times 6 \times 10^{-3})^2}$$

$$\therefore \quad \boxed{Z_1 = 37.72 \ \Omega}$$

$$\therefore \qquad I_1 = \frac{V_1}{I_1} = \frac{94.18}{37.72} = 2.496 \text{ Amp}$$

$$\therefore \qquad \boxed{I_1 = 2.496 \text{ Amp}}$$

Similarly,

$$I_2 = \frac{2 \times 220}{2 \cdot \sqrt{2}\, \pi} \frac{\sin(2\pi \times 0.4)}{Z_1}$$

$$\boxed{I_2 = 0.7716 \text{ Amp}}$$

and $\quad I_3 = \dfrac{2 \times 220}{3\sqrt{2}\, \pi} \dfrac{\sin(3\pi \times 0.4)}{Z_1} = -\dfrac{0.5144}{3} = -0.1714 \text{ Amp}$

$$\therefore \qquad \boxed{I_3 = -0.1714 \text{ Amp}}$$

(f) Average supply current is

$$I_{TAV} = \frac{\alpha\,(V_S - E)}{R} - \frac{L}{RT}\,(I_{mx} - I_{mn})$$

$$= \frac{0.4\,(220 - 44)}{1.5} - \frac{6}{1.5 \times 1}\,(33.72 - 24.974)$$

$$= 46.93 - 34.984$$

$$\boxed{I_{TAV} = 11.943 \text{ Amp}}$$

(g) $\qquad$ Input power $= V_S \times$ average supply current
$$= 220 \times 11.943$$

$$\boxed{P_i = 2.62746 \text{ kW} = 2.62746 \text{ kW}}$$

Power absorbed by load e.m.f. $= E \times I_o$
$$= 44 \times 29.33 = 1.290 \text{ kW}$$
Power loss in resistor, $R = 2.62746 - 1.29052$
$$= 1.287 \text{ kW}$$

(h) $\qquad I_{r.m.s.} = \sqrt{I_{ave}^2 + I_1^2 + I_1^2 + I_3^2}$

$$= \sqrt{(29.33)^2 + (2.496)^2 + (0.7719)^2 + (0.1716)^2}$$

$$= \sqrt{867.104} = 29.446 \text{ Amp}$$

$$\therefore \qquad \boxed{I_{r.m.s.} = 29.446 \text{ Amp}}$$

(i) Power loss in resistor = $I^2 R$ = 1.287 kW

$$\therefore \qquad I_{r.m.s.} = \sqrt{\frac{1287}{1.5}}$$

$$\boxed{I_{r.m.s.} = 29.29 \text{ Amp}}$$

Problem 5.11: A voltage-commutated chopper delivers power to RLE load for which $R = 0$ and $L = 8$ mH. For a chopping frequency of 200 Hz and d.c. source voltage of 400 V, find the chopper duty cycle dc so as to limit the load current excursion to 40 A.

Solution:

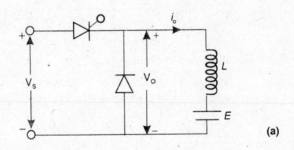

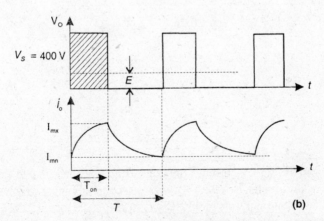

Fig. P. 5.11 (a) and (b)

Let the duty cycle be α to limit the charging load current excursion to 40 A.

The average output voltage is given by

$$V_o = \alpha V_S$$

As the average value of voltage drop across inductor is zero, hence:

$$\therefore \quad E = V_o = \alpha V_S$$

During T_{on}, the difference in source voltage V_S and load e.m.f. E, i.e. $(V_S - E)$ appears across L.

$\therefore$ During T_{on}, volt time are applied to inductor
$$= (400 - \alpha\, 400)\, T_{on} = 400\, T_{on}\, (1 - \alpha) \text{ V-sec}$$

Also during T_{on}, the current through L rises from I_{mn} to I_{mx}. From this, voltage time area across L during this current change is given by

$$\int_0^{T_{on}} V_L \cdot dt = \int_0^{T_{on}} L\frac{di}{dt}\cdot dt = \int_{I_{mn}}^{I_{mx}} L \cdot d_i = (I_{mx} - I_{mn})\, L = L\, \Delta I$$

These two volt time areas during T_{on} must be equal

$$\therefore \quad 400\, T_{on}\, (1 - \alpha) = L\, \dot{\Delta}\, I$$
$$\text{or} \quad 400\, \alpha\, T\, (1 - \alpha) = L\, \dot{\Delta}\, I$$

$$\text{or} \quad \alpha\,(1 - \alpha) = \frac{8 \times 10^{-3} \times 40}{400 \times T}$$

$$\text{or} \quad \alpha - \alpha^2 = 8 \times 10^{-4} \times 200 = 0.16$$

$$\text{or} \quad \alpha^2 - \alpha + 0.16 = 0$$

$$\text{or} \quad \alpha = \frac{-(-1) \pm \sqrt{(-1)^2 - 4 \times 0.16}}{2 \times 1}$$

$$\alpha = \frac{1 \pm 0.6}{2}$$

$$\boxed{\alpha = 0.8 \text{ and } 0.2}$$

Problem 5.12: A current commutated chopper has the following data : source voltage = 220 V d.c.; peak commutating current = 1.8 times the load current; main SCR t_q = 20 μs; factor of safety = 2; load current = 180 A. Determine the values of the commutating inductor and capacitor, maximum capacitor voltage and the peak commutating current.

Solution:

$$\because \qquad x = \frac{I_{cp}}{I_o} = 1.8, \; t_q = 20 \; \mu \; \text{sec}$$

$$\therefore \qquad t_C = t_q + \Delta t, \text{ since factor of safety} = 2$$

$$\therefore \qquad \Delta t = 20 \; \mu \; \text{sec}$$

$$\therefore \qquad t_C = 20 + 20 = 40 \; \mu \; \text{sec}$$

$$L = \frac{V_S \cdot t_C}{x \, I_O \left[\pi - 2\sin^{-1}\left(\dfrac{1}{x}\right) \right]}$$

or

$$L = \frac{220 \times 40 \times 10^{-6}}{1.8 \times 180 \left[\pi - 2 \times \sin^{-1}\left(\dfrac{1}{1.8}\right) \times \dfrac{\pi}{180} \right]}$$

$$L = \frac{220 \times 40 \times 10^{-6}}{1.8 \times 180 \times 1.9635}$$

$$L = 13.88 \times 10^{-6}$$

or

$$\boxed{L = 13.83 \; \mu\text{H}}$$

$$C = \frac{x \, I_o \cdot t_C}{V_S \left[\pi - 2\sin^{-1}\left(\dfrac{1}{x}\right) \right]} = \frac{1.8 \times 180 \times 40 \times 10^{-6}}{220 \left[\pi - 2\sin^{-1}\left(\dfrac{1}{6}\right) \dfrac{\pi}{180} \right]}$$

$$= \frac{1.8 \times 180 \times 40 \times 10^{-6}}{220 \times 1.9635} = 30.002 \times 10^{-6}$$

or

$$\boxed{C = 30.002 \; \mu\text{F}}$$

Peak capacitor voltage is

$$V_{CP} = 220 + 180 \sqrt{\frac{13.83}{30.002}} = 342.21 \; \text{V}$$

$$\therefore \qquad \boxed{V_{CP} = 342.21 \; \text{V}}$$

Peak commutating current

$$I_{CP} = x \, I_0 = 1.8 \times 180$$

$$\boxed{I_{CP} = 324 \; \text{Amp}}$$

Problem 5.13: The commutating components for a current-commutated chopper are $C = 40$ μF and $L = 18$ μH. D.C. source voltage is 220 V and load current is constant at a value of 180 A during the commutation interval. For this chopper, calculate :
(a) Circuit turn-off time for main thyristor.
(b) Circuit turn-off time for auxiliary thyristor.
(c) Total commutation interval.

Solution: $I_{CP} = V_S \sqrt{\dfrac{C}{L}} = 220 \sqrt{\dfrac{40}{18}} = 327.95$ Amp

$\therefore$ $x = \dfrac{I_{CP}}{I_o} = \dfrac{327.95}{180} = 1.822$

$\therefore$ $\theta_1 = \sin^{-1}\left(\dfrac{1}{x}\right) = 33.288°$

(a) Circuit turn-off time for main thyristor

$t_4 - t_3 = t_C = \left[\pi - 2\sin^{-1}\left(\dfrac{1}{x}\right)\right] \sqrt{LC}$

$= [\pi - 1.161] \sqrt{40 \times 18 \times \left(10^{-6}\right)^2} = 1.97\sqrt{720} \times 10^{-6}$

$\boxed{t_C = 53.118 \text{ μ sec}}$

(b) Circuit turn-off time for auxiliary thyristor

$t_4 - t_2 = t_{C_1} = \left[\pi - \sin^{-1}\left(\dfrac{1}{x}\right)\right] \sqrt{LC}$

$= [\pi - 0.5805] \sqrt{40 \times 18 \times \left(10^{-6}\right)^2} = 2.56 \times \sqrt{720} \times 10^{-6}$

$\boxed{t_{C_1} = 68.72 \text{ μ sec}}$

(c) Total commutation interval is

$(t_6 - t_1) = \left(\dfrac{5\pi}{2} - \theta_1\right) \sqrt{LC} + 2CV_S \cdot \dfrac{\sin^2\theta\dfrac{1}{2}}{I_0}$

or $(t_6 - t_1) = (7.27) \sqrt{40 \times 18} \times 10^{-6} + 2 \times 40 \times 10^{-6} \times$

$$220 \frac{\sin^2\left(\dfrac{33.288}{2}\right)}{180}$$

$$= 195.15 \times 10^{-6} + 8.02 \times 10^{-6}$$

$$\boxed{(t_6 - t_1) = 203.17 \ \mu \ sec}$$

Problem 5.14: A load commutated chopper, fed from 230 V d.c. source, has a constant load current of 50 A. For a duty cycle of 0.4 and a chopping frequency of 2 kHz, compute.
(1) The average output voltage.
(2) The value of commutating capacitance.
(3) Total commutation interval.

Solution: (1) Average output voltage

$$V_o = \alpha \ V_S = 0.4 \times 230 = 92 \ V$$

∴ $\boxed{V_o = 92 \ V}$

(2) $$C = \frac{I_o \ T_{on}}{2 V_S} = \frac{I_o \cdot \alpha T}{2 V_S} = \frac{50 \times 0.4}{2 \times 230} \times \frac{1}{2 \times 10^3}$$

or $C = 0.02173 \times 10^{-3}$

or $\boxed{C = 21.73 \ \mu F}$

(3) Total commutation interval

$$= T_{on} = \frac{2 C V_S}{I_o} = \frac{2 \times 21.73 \times 10^{-6} \times 230}{50}$$

$$\boxed{T_{on} = 200 \ \mu \ sec}$$

(4) Circut turn-off time for one thyristor pair

$$t_C = \frac{1}{2} \ T_{on} = \frac{200}{2} \ \mu \ sec = 100 \ \mu \ sec$$

∴ $\boxed{t_C = 100 \ \mu \ sec}$

Problem 5.15: A separately excited d.c. motor has the following name plate data : 220 V, 100 A, 2200 r.p.m. The armature resistance is 0.1 Ω and in ductance is 5 mH. The motor is fed by a chopper which is operating from a d.c. supply of 250 V. Due to restrictions in the power circuit, the chopper can be operated over a duty cycle range from 20% to 80%. Determine the range of speeds over which the motor can be operated at rated torque.

Solution: For a separately excited d.c. motor back e.m.f. is given as

$$E_b = V_t - I_a r_a = 220 - 100 \times 0.1 = 210 \text{ V}$$

Since $\quad E_b = K_a \phi \omega_n$

or $\quad K_a \phi = \dfrac{E_b}{\omega_n} = \dfrac{210}{2200} = 0.095 \, \dfrac{V}{\text{r.p.m.}}$

d.c. supply voltage = 250 V

Duty cycle $\alpha = \dfrac{20}{100} = 0.2$ and $\dfrac{80}{100} = 0.8$

let $\quad \alpha_1 = 0.2$ and $\alpha_2 = 0.8$

∴ Applied voltage to motor for α_1

$$V_{a_1} = 250 \times 0.2 = 50 \text{ V}$$

and for α_2, the applied voltage is

$$V_{a_2} = 250 \times 0.8 = 200 \text{ V}$$

back e.m.f.

$$E_{b_1} = V_{a_1} - I_a r_a = 50 - 100 \times 0.1 = 40 \text{ V}$$

and $\quad E_{b_2} = V_{a_2} - I_a r_a = 200 - 100 \times 0.1 = 190 \text{ V}$

∴ $E_b \propto \omega_S$ (Speed)

∴ Speed at $E_{b_1} = \dfrac{40}{K_a \phi} = \dfrac{40}{0.095} \approx 400$ r.p.m.

and speed at $E_{b_2} = \dfrac{190}{K_a \phi} = \dfrac{190}{0.095} \approx 1900$ r.p.m.

∴ Range of speed is 400 r.p.m. to 1900 r.p.m.

Problem 5.16: An ideal chopper operating at a frequency of 500 Hz feeds an R-L load having $R = 30$ Ω and $L = 9$ mH from a 48 V battery.

The load is shunted by a free wheeling diode. Battery is loss less. Assuming duty cycle of chopper to be 50% compute (a) peak load current, (b) minimum load current, (c) average load current, (d) average load voltage, and (e) current exersion in load current.

Solution: Output voltage of a chopper is given by

$$V_o = \alpha\, V_S, \qquad \text{where } \alpha = \text{duty cycle}$$
$$= 0.5 \times 48 = 24 \text{ V}$$

$$\therefore \qquad \boxed{V_o = 24 \text{ V}}$$

$$T_{on} = T_{off} = \frac{24}{48 \times 500} = 1\text{m sec}$$

(a) Peak load current

$$I_{max} = \frac{V_S}{R} \frac{\left(1 - e^{-\frac{t_{on}}{T_a}}\right)}{\left(1 - e^{-\left(t_{on} + \frac{t_{off}}{T_a}\right)}\right)}$$

$$t_1 = t_2 = T_{on} = 1 \text{ m sec}$$

Time constant $T = \dfrac{L}{R} = \dfrac{9 \times 10^3}{30} = 300\ \mu s = 0.3 \text{ ms}$

$$\therefore \qquad I_{max} = \frac{48}{30} \frac{\left(1 - e^{-\frac{1}{0.3}}\right)}{\left(1 - e^{-\frac{2}{0.3}}\right)} = 1.6\, \frac{(1 - 0.036)}{(1 - 0.001)}$$

$$\boxed{I_{max} = 1.544 \text{ Amp}}$$

(b) $\qquad I_{min} = I_{max}\, e^{-\frac{t_2}{T}} = 1.544 \times 0.036$

$$\boxed{I_{min} = 0.056 \text{ Amp}}$$

(c) Average current

$$I_o = \frac{24}{30} = 0.8 \text{ Amp}$$

(d) Average voltage $(V_o) = 24$ V

(e) Current excussion in load current

$$= 1.544 - 0.056 = 1.488 \text{ Amp}$$

Problem 5.17: A seperately excited d.c. motor is fed from a chopper operating at 500 Hz with a duty cycle of 50% and as drawing an average current of 10 A from a 200 V d.c. source. A field wheeling diode is connected across it. The motor has negligible armature resistance, a field inductance of 50 mH and a torque constant of 0.5 N.m/A. Determine the minimum and maximum motor current, motor back e.m.f. and the mechanical torque developed.

Solution: Separately excited d.c. motor is fed from a chopper and average output current

$$I_o = 10 \text{ A, field inductance} = 50 \text{ mH}$$

Torque constant = 0.5 N-m/A; R_a is negligible small

$$I_{max} = \frac{V_S}{r_a} \left[\frac{1 - e^{\frac{t_{on}}{T_a}}}{1 - e^{\frac{t_{on} + t_{off}}{T_a}}} \right] - \left\{ \frac{K_a \phi N}{r_a} \right\} = \frac{200}{r_a} \left[\frac{1 - e^{\frac{0.5}{T_a}}}{1 - e^{\frac{2 \, ms}{T_a}}} \right] - \left\{ \frac{K_a \phi N}{r_a} \right\}$$

Here $T_a = \dfrac{L}{r}$, and $T = T_{on} + T_{off} = \dfrac{1}{500} = 2$ ms

With neglible resistance I_{max} and

$$I_{min} = \frac{V_S}{r_a} \left\{ \frac{e^{\frac{t_{on}}{T_a} - 1}}{e^{\frac{T}{T_a}} - 1} \right\} - \frac{E}{r_a} \quad \text{can not be determine.}$$

Back e.m.f. $E_b = V_t - I_a r_a \approx V_t = 220$ V (as r_a is very small.)

Mechanical torque developed

$$T = K_a \phi I_a = 0.5 \times 10 = 5 \text{ N-m}$$

Problem 5.18: A chopped feeds an R-L load comprising a resistance of 10 Ω and an inductance of 15 mH. Chopper frequency is 500 Hz and it is fed from α 220 V dc and α = 50 %. Determine :

(a) Maximum and minimum values of load current.
(b) Peak to peak ripple.
(c) Average value of load current.

Solution: Source voltage V_S = 220 V

Chopper frequency = 500 Hz

∴ Chopper period = 2 ms

$$T_a = \frac{L}{R} = \frac{15\,\text{mH}}{10} = 1.5 \text{ ms}$$

T_{on} = 1 ms and T_{off} = 1 ms.

(a) ∴

$$I_{max} = \frac{V_S}{R}\frac{\left(1 - e^{-\frac{T_{on}}{T_a}}\right)}{\left(1 - e^{-\frac{T}{T_a}}\right)} = \frac{220}{10}\frac{\left(1 - e^{-\frac{1}{1.5}}\right)}{\left(1 - e^{-\frac{2}{1.5}}\right)}$$

$$\boxed{I_{max} = 14.5490 \text{ Amp}}$$

and

$$I_{min} = \frac{V_S}{R}\frac{\left(e^{\frac{T_{on}}{T_a}} - 1\right)}{\left(e^{\frac{T}{T_a}} - 1\right)} = \frac{220}{10}\frac{\left(e^{-\frac{1}{1.5}} - 1\right)}{\left(e^{-\frac{2}{1.5}} - 1\right)}$$

$$\boxed{I_{min} = 7.4697 \text{ Amp}}$$

(b) Peak to peak ripple = 7.0793 Amp

(c) Average value of load current = $\dfrac{220 \times 0.5}{10}$ = Amp

---

Problem 5.19: A chopper feeds an R-L load. The inductance is very large and the load current is ripple free. Current may be assumed to be constant during commutation. The chopper is fed from a 5000 V supply and average value of load current is 80 A. The main thyristor has a turn off time of 15 μs.

(a) Neglecting the effect of inductance inseries with auxiliary thyristor, determine the value of capacitance in the commutation circuit.
(b) If the rate of change of auxiliary thyristor current is to be limited to 50 A/µs, determine the value of protective inductance.
(c) The recharging process takes a time of 75 µs. Determine the value of inductance of the recharging circuit.
(d) What is the peak value of thyristor current during recharging process?

Solution: (a) If zero current is maintained for a time equal to turn-off time then a thyristor is successfully turned off, i.e. the commutation is achieved. It can be possible by applying a capacitor voltage in reverse direction. This negative voltage becomes zero due to discharge of the capacitor. The time is called circuit turn-off time. The minimum value of circuit turn-off time should be greater than and equal to circuit turn-off time. So,

$$t_C = \frac{C \cdot V_S}{I_o} \geq t_q$$

$$\therefore \qquad \boxed{C = \geq \frac{I_o \cdot t_q}{V_S}}$$

or $$C = \frac{80 \times 15 \times 10^{-6}}{500} = 2.4 \ \mu F$$

Minimum value of capacitance = 2.4 µF.
Normally to have safe commutation

$$t_C = 2.5 \ t_q$$

$$\therefore \qquad C = 2.5 \times 2.4 = 6 \ \mu F$$

(b) The rate of charge of current in the auxiliary thyristor is 50 $\dfrac{A}{\mu s}$.

$$\therefore \qquad L_1 \frac{di}{dt} = 500 \ V$$

or $$L_1 = \frac{500 \ V}{50 \dfrac{A}{\mu s}} = 10 \ \mu H$$

(c) The frequency of oscillation of recharging

$$\text{Circuit} = \frac{1}{2\pi \sqrt{L_2 C}}$$

$$\text{Period} = 2\pi \sqrt{L_2 C}$$

The recharging is complete in half of the period.

$\therefore$ $\qquad 75 = \pi \sqrt{L_2 C}$

$\therefore$ $\qquad \boxed{L_2 = 237.7 \ \mu\text{H}}$

With 6 μF capacitor

$$\boxed{L_2 = \frac{75}{\pi^2 \times 6} = 95 \ \mu\text{H}}$$

(d) Peak value of thyristor current = 75 + peak value of recharging current.

$\therefore$ $\qquad \dfrac{1}{2} L i^2 = \dfrac{1}{2} C V^2$

$\therefore$ $\qquad i = 500 \sqrt{\dfrac{2.4}{237.7}} = 50.25$ A

$\therefore$ Peak value of current through the main thyristor is

$\qquad 75 + 50.26 = 125.6$ A $\qquad$ with 2.4 μF and $L_2 = 23.7.7 \ \mu$H
and similarly
$\qquad 75 + 125.5 = 200.5$ A $\qquad\qquad$ with 6 μF and $L_2 = 95 \ \mu$H

Problem 5.20: A separately excited d.c. motor rated at 37.5 kW takes an armature current of 180.5 A from a 230 V supply. Its rated speed is 1300 r.p.m. The armature resistance is 0.05 Ω. The motor is controlled from a d.c. chopper for speed control in the range 300-1200 r.p.m. at constant load torque. The chopper operates on a variable frequency with ON time of 4μ sec. Determine the time ratios of chopper.

Solution: Back e.m.f. of the motor at 1300 r.p.m.

$$E_b = V_t - I_d r_a = 230 - 180.5 \times 0.05$$

$$\boxed{E_b = 221 \text{ V}}$$

The armature current remains the same at 300 and 1200 r.p.m. The induced voltage at these speeds are

$$= 221 \times \frac{300}{1300} = 51 \text{ V}$$

and $= 221 \times \frac{1200}{1300} = 204$ V respectively.

Supply voltage = 60.03 V 213.03 V

Time ratio = 0.261 0.9262

T = 15.33 ms 4.319 m sec

f = 65 Hz 231.5 Hz

CHAPTER 6

INVERTERS

A device that converts d.c. power into a.c. power at desire output voltage and frequency is called an inverter. It can be divided into two types, one is voltage source inverter and other is current source inverters. A voltage source inverter (VSI) is one in which the d.c. source has small or negligible impedance. A current source inverter (CSI) is one in which current is fed from a d.c. source of high impedance. In CSI, output current are not affected by the load.

Inverters are widely used for adjustable-speed a.c. drives, induction heating, UPS, HVDC transmission lines, stand by air craft power supplies etc.

Voltage source inverter

(1) $1 - \phi$ bridge inverter are classified into two types (a) $1 - \phi$ half bridge (b) $1 - \phi$ full bridge inverters.

(a) $1 - \phi$ half bridge :

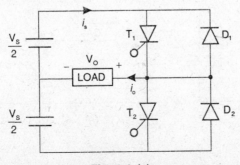

Fig. 6.1 (a)

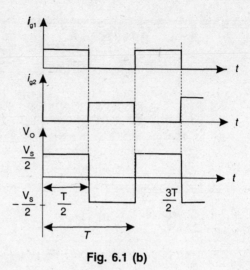

Fig. 6.1 (b)

During $0 < t \le \dfrac{T}{2}$, thyristor T_1 conducts and the load is subjected to a voltage $\dfrac{V_S}{2}$ due to the upper voltage source. During $\dfrac{T}{2} < t \le T,$ thyristor T_2 conducts and the load is subjected to a voltage $\left(-\dfrac{V_S}{2}\right)$ due to the lower voltage source. Thus it is seen that output voltage is an alternating voltage.

The main draw back of half bridge inverter is that it requires 3-wire d.c. supply.

(b) 1 – ϕ full bridge inverter :

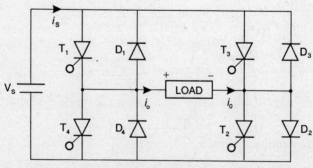

Fig. 6.2 (a)

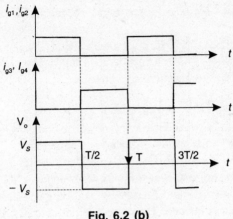

Fig. 6.2 (b)

During $0 < t \leq \dfrac{T}{2}$, thyristor T_1 and T_2 conducts and the load is subjected to a voltage V_S. During $\dfrac{T}{2} < t \leq T$, thyristor T_3 and T_4 conducts and the load is subjected to a voltage $-V_S$. Thus it is seen that output voltage is an alternating voltage.

For both $1-\phi$ half bridge and $1-\phi$ full bridge inverter with RL load the load current i_o is not in phase with voltage V_o. So we connect diodes in antiparallel with thyristors which will allow the current to flow when the main thyristor are turned off. These diode is known as feedback diodes.

Fourier Analysis of $1 - \phi$ inverter output voltage

The output voltage waveforms do not depend on the nature of load. For $1 - \phi$ half-bridge inverter.

$$V_o = \sum_{n=1,3,5}^{\infty} \frac{2V_S}{n\pi} \sin n\omega t \text{ Volts}$$

For $1 - \phi$ full bridge inverter :

$$V_o = \sum_{n=1,3,5}^{\infty} \frac{4V_S}{n\pi} \sin n\omega t \text{ Volts}$$

Inverter commutation

In voltage source inverters, thyristor remain forward biased by the d.c. supply voltage. This entails the use of forced commutation for inverter circuit. Three different types of commutation are classified as :

   (a) Modified *MC* Murray half-bridge Inverter.
   (b) Modified *MC* Murray-Bedford half-bridge inverter.
   (c) Auxiliary commutating supply.

3 – φ Bridge Inverter

For providing adjustable frequency power to industrial application, 3 – φ inverter are more common than 1 – φ inverters.

   A basic 3 – φ inverter is a six-step bridge inverter. It uses a minimum of six thyristors. For one cycle of 360°, each step would be of 60° interval for a six-step inverter and thyristor would be gated at regular intervals of 60° in proper sequence. There are two possible patterns of gating the thyristors. In one pattern each thyristor conducts for 180° and in the other, each thyristor conducts for 120°. But in both of these patterns gating signals are applied removed at 60° intervals of the output voltage waveform.

Voltage control in 1 – φ inverters The various methods are:

   (a) External control of a.c. output voltage.
   (b) External control of d.c. input voltage.
   (c) Internal control of a.c. inverter.

(a) External control of a.c. output voltage

   (1) *AC voltage control* : In this method a.c. voltage controller is inserted between inverter and the load terminals. Due to harmonic, this method is rarely used.
   (2) *Series inverter control* : This method of voltage control involves the use of two or more inverters in series.

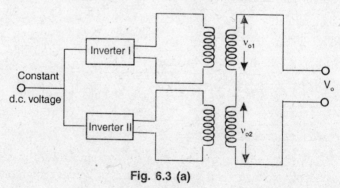

Fig. 6.3 (a)

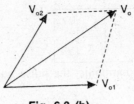

Fig. 6.3 (b)

The output voltage V_o is given as :

$$V_o = \left[V_{o_1}^2 + V_{o_2}^2 + 2V_{o_1} \cdot V_{o_2} \cos\theta\right]^{\frac{1}{2}}$$

where $\quad\quad \theta$ = Phase angle difference between V_{o_1} and V_{o_2} and it depends upon the firing angle of two inverters.

(b) External control of d.c. input voltage

In this techniques, d.c. voltage input to inverter is controlled by means of components external to the inverter.

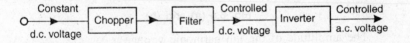

(c) Internal control of inverter

Output voltage from an inverter can also be adjusted by exercising a control with in the inverter itself. The most efficient method is pulse-width control method.

(1) *Single pulse modulation*

In this modulation the range of pulse width '$2d$' varies from 0 to π. The output voltage is controlled by varying the pulse width $2d$.

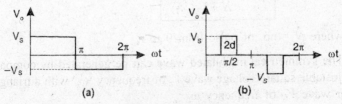

Fig. 6.4 (a) and (b)

The r.m.s. value of output voltage is

$$V_{or} = V_s \left[\frac{2d}{\pi} \right]^{1/2}$$

For reducing nth harmonic, the pulse width '$2d$' should be made equal to $\frac{2\pi}{n}$, i.e.

$$2d = \frac{2\pi}{n}$$

(2) *Multiple pulse modulation*

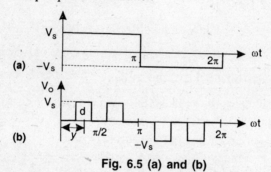

Fig. 6.5 (a) and (b)

The amplitude of the nth harmonic voltage is :

$$V_n = \frac{8V_S}{n\pi} \sin ny \cdot \sin \frac{nd}{2}$$

This shows that the magnitude of V_n depends upon γ and d.

The above expression also shows that when $\gamma = \frac{\pi}{n}$ or $d = \frac{2\pi}{n}$, nth harmonics can be eliminated from the output voltage. In general

$$\gamma = \frac{\pi - 2d}{N+1} + \frac{d}{N}$$

where N = no. of pulse from 0 to π.

The symmetrical modulated wave can be generated by comparing an adjustable square voltage wave V_r of frequency 'ω_r' with a triangular carrier wave V_C of frequency ω_c.

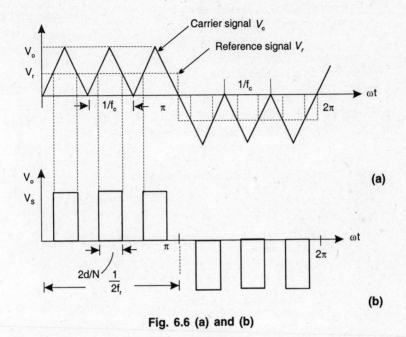

(a)

(b)

Fig. 6.6 (a) and (b)

No. of pulse per half cycle $= \dfrac{\dfrac{1}{2}f_r}{\dfrac{1}{f_C}}$

or $\qquad N = \dfrac{\dfrac{1}{2}f_r}{\dfrac{1}{f_C}} = \dfrac{f_C}{2f_r} = \dfrac{\omega_C}{2\omega_r}$

the pulse width is

$$\dfrac{2d}{N} = \left(\dfrac{\pi}{N} - \dfrac{\pi}{N} \dfrac{V_r}{V_C} \right)$$

$$\boxed{\dfrac{2d}{N} = \left(1 - \dfrac{V_r}{V_C} \right) \dfrac{\pi}{N}}$$

In MPM method, lower order harmonics can be eliminated by a proper choice of $2d$ and γ. But the r.m.s. voltage is the same.

$$V_{or} = V_S \left[\frac{2d}{\pi} \right]^{\frac{1}{2}}$$

(3) *Sinusoidal-pulse modulation*

In MPM method, the pulse width is equal for all the pulses. But in sinusoidal-pulse modulation, the pulse width is a sinusoidal function of the angular position of the pulse in a cycle.

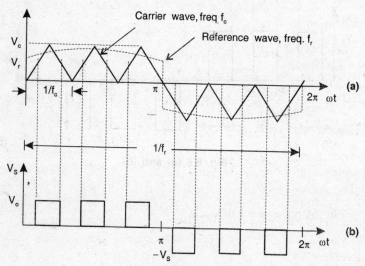

Fig. 6.7 (a) and (b)

No. of pulse per half cycle $(N) = \dfrac{f_C}{2 f_r}$

$\therefore$

$$N = \frac{f_C}{2 f_r}$$

The magnitude of fundamental component of output voltage is proportional to modulation index $\left(\dfrac{V_r}{V_C} \right)$, but modulation index can never be more than unity. Thus the output voltage is controlled by varying modulation index.

Current source inverter

In the voltage source inverter, input voltage is maintained constant and the amplitude of output voltage does not depend on the load. However, the magnitude of load current depends upon the nature of the load impedance.

In the current source inverter (CSI) input current is constant but adjustable. The amplitude of output current is independent of the load impedance. A CSI does not require any feedback diodes as in the voltage source inverter.

Application of CSI

    (1) Induction heating.
    (2) Speed control of a.c. motors.
    (3) Synchronous motor starting.
    (4) Lagging VAr compensation.

Series inverter

Inverter in which commutating components are permanently connected in series with the load are called series inverter. The series circuit so formed must be underdamped. As the current attains zero value due to the nature of the series circuit, series inverter are also called self commutated or load commutated inverter. These inverters operate at high frequency (200 Hz to 100 kHz). It is widely used in induction heating, fluorescent lighting etc.

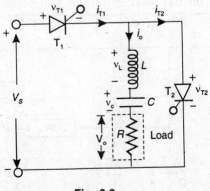

Fig. 6.8

Parallel inverter

In parallel inverters, the commutating capacitor is connected in parallel with the load. When the capacitor applies a reverse potential across the conducting SCR, commutation is achieved. This is also referred as voltage commutation.

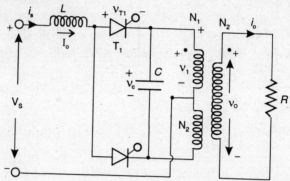

Fig 6.9. 1 – ϕ capacitor commutated parallel inverter

PROBLEMS

Problem 6.1: A single-phase full-bridge inverter may be connected to a load consisting of (a) *R*, (b) RL or RLC over damped, (c) RLC underdamped. For all these loads draw the load voltage and load current wave forms under steady operating condition. Discuss the nature of these wave forms. Also indicate the conduction of the various elements of the inverter circuit.

Is it possible for this inverter to have load commutation? Explain.

Solution:

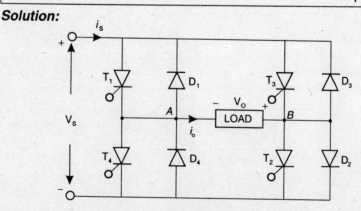

Fig. P. 6.1

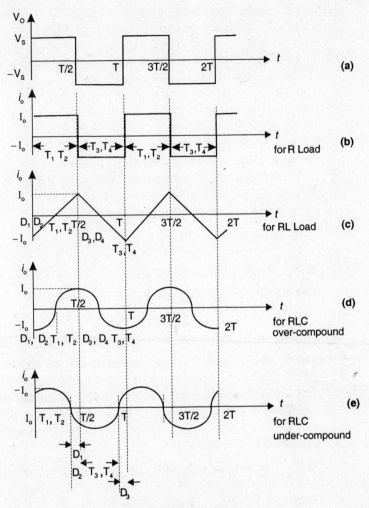

Fig. P. 6.1 (a) - (e)

Under steady : State condition load voltage waveform does not depends on the nature of load. The load voltage is given by :

$$V_o = \begin{cases} V_S, & 0 < t < \dfrac{T}{2} \\[2mm] -V_S, & \dfrac{T}{2} < t < T \end{cases}$$

However, the load current is dependent upon the nature of load.

(A) *For R Load* : For resistive load, load current wave form i_o is identical with load voltage wave form V_o and diodes $D_1 - D_4$ do not come into conduction.

(B) *For RL and RLC over damped loads* : The load current wave forms for RL and RLC over damped loads are shown in Figs (d) and (e) respectively. Before $t = 0$, thyristors T_3, T_4 are conducting and load current i_o is flowing from B to A, i.e. in the reverse direction. After T_3, T_4 are turned off at $t = 0$, current i_o can not change its direction immediately because of the nature of load. As a result, diodes D_1, D_2 start coducting after $t = 0$ and allow i_o to flow against the supply voltage V_S. As soon as D_1, D_2 begin to conducting, load is subjected to V_S. Untill the diode current does not falls to zero, SCR T_1 and T_2 does not turn on. At gated period of $\dfrac{T}{4}$ sec, SCR T_1 and T_2 start conducting till gating period of $\dfrac{T}{2}$. After that T_1, T_2 are turned off by forced commutation and as load current can not reverse immediately, diodes D_3 and D_4 come into conduction to allow the flow of current i_o after $\dfrac{T}{2}$. When D_3, D_4 diode current drops to zero; T_3, T_4 are turned on as these are already gated.

(C) *RLC under damped load* : For RLC under damped load no commutation circuitry is needed, because SCR T_1, T_2 will get commutated naturally when $\left(\dfrac{T}{2} - t_1\right) > t_q$. The load current i_o for RLC under damped load is shown in Fig. (*f*). After $t = 0$, T_1, T_2 are conducting the load current. As i_o through T_1, T_2 reduces to zero at t_1, these SCR are turned off before T_3, T_4 are gated. As T_1, T_2 stop conducting current through the load reverses and is now carried by diodes D_1, D_2 as T_3, T_4 are not yet gated. The diodes D_1, D_2 are connected in antiparallel to T_1, T_2 ; the voltage drop in these diodes appears as a reverse bias across T_1, T_2. If duration of this reverse bias is more than the SCR turn-off time t_q, i.e. if; $\left(\dfrac{T}{2} - t_1\right) > t_q$ T_1, T_2 will get commutated naturally and therefore no commutation circuitry, will be needed. This method of commutation, known as load commutation, is in fact uses in high frequency inverters used for induction heating.

So, for RLC under damped load $1 - \phi$ full inverter, it is possible to have load commutation, when duration of reverse bias is more than SCR turn-of time t_q, i.e. if $\left(\dfrac{T}{2} - t_1\right) > t_q$.

Problem 6.2: For a $1 - \phi$ full bridge inverter, $V_S = 230$ V d.c., $T = 1$ ms. The load consists of RLC in series with $R = 1.2\ \Omega$, $\omega L = 8\ \Omega$, $\dfrac{1}{\omega c} = 7\ \Omega$.

(a) Sketch the waveforms for load voltage V_o, fundamental component of output current i_{o1}, source current i_s and voltage across thyristor 1. Indicate the devices that conduct during different intervals of one cycle. Find also the r.m.s. value of fundamental component of load current.

(b) Find the power delivered to load due to fundamental component.

(c) Check whether forced commutation is required or not.

Solution: (a) The load voltage wave form V_o, fundamental component of output current i_{o1}, source current i_s and voltage across thyristor (V_{T_1}) are shown as :

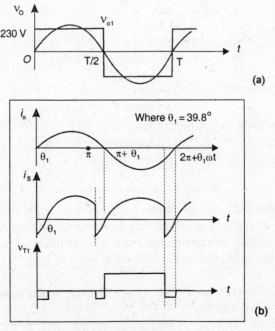

Fig P. 6.2 (a)-(b)

Waveforms of V_o, i_o, i_S and V_{T1}.
r.m.s. value of load voltage is :

$$V_{o1} = \frac{4 V_S}{\pi\sqrt{2}} = \frac{4 \times 230}{\pi\sqrt{2}} = 207.1 \text{ V}$$

∴ r.m.s. value of current

$$I_{o1} = \frac{V_{o1}}{z_1} = \frac{V_{o1}}{\left[R^2 + \left(\omega L - \dfrac{1}{\omega C}\right)^2\right]^{\frac{1}{2}}} = \frac{207.1}{\sqrt{(1.2)^2 + (8-7)^2}}$$

$$= \frac{207.1}{\sqrt{2.44}} = \frac{207.1}{1.562}$$

$$\boxed{I_{o1} = 132.586 \text{ Amp}}$$

The fundamental cmponent of current as a function of time is

$$i_{o1} = \sqrt{2} \, I_{o1} \sin(\omega t - \phi_1)$$

where $\phi_1 = \tan^{-1} \dfrac{X_L - X_C}{R} = \tan^{-1} \dfrac{8-7}{1.2}$

$$\boxed{\phi_1 = 39.80^\circ}$$

∴ $i_{o1} = 187.5 \sin(\omega t - 39.80^\circ)$

(b) Power delivered to load $= I_{o1}^2 \, R$

$$= (132.586)^2 \times 1.2 = 21094.85 \text{ W}$$

(c) As the load current i_o does not change from positive to negative at an angle $\omega t < \pi$, no time is available for SCR to turn-off. Hence force commutation is required for this circuit.

Problem 6.3: A single phase bridge inverter is fed from 230 V d.c. In the output voltage wave only fundamental component of voltage is considered. Determine the r.m.s. current rating of an SCR and a diode of the bridge for the following types of loads (a) $R = 2 \ \Omega$ (b) $\omega L = 2 \ \Omega$.

Find also the repetitive peak voltage that may appear across a thyristor in parts (a), and (b).

Solution: The load current i_o can be expressed as

$$i_o = \sum_{n=1,3,5}^{\infty} \frac{4V_S}{n\pi z_n} \sin(n\omega t - \phi_n) \text{ Amp.}$$

For fundamental component of load current $n = 1$ and for resistive load $\phi_n - 0°$.

$$\therefore \qquad i_o = \frac{4V_S}{\pi z_n} \sin\omega t$$

So, $\qquad i_m = \dfrac{4V_S}{\pi z_n}$

(a) For $\qquad R = 2\ \Omega$

$$i_m = \frac{4 \times 230}{\pi \times 2} = 146.422 \text{ Amp}$$

The r.m.s. value of thyristor current is :

$$i_T = \left[\frac{1}{2\pi} \int_0^{\pi} \left(I_m \sin\omega t \right)^2 d(\omega t) \right]^{\frac{1}{2}}$$

$$i_T = \frac{I_m}{2} = \frac{146.422}{2}$$

$$\therefore \qquad \boxed{i_T = 73.211 \text{ Amp}}$$

For resistive load diode does not conduct so diode current $i_{D_1} = 0$.

$$\therefore \qquad \boxed{i_{D_1} = 0}$$

Repetitive peak voltage across thyristor $= V_S = 230$ V

(b) For $\omega L = 2\ \Omega$ $\left[\text{For inductive load, load current conducts for} \right.$

0 to $\dfrac{\pi}{2}\Big]$

$$\boxed{i_m = \frac{4V_S}{\pi Z_n} = \frac{4 \times 230}{\pi \times 2} = 146.422 \text{ Amp}}$$

The r.m.s. value of thyristor current is

$$i_{T_1} = \left[\frac{1}{2\pi} \int_0^{\frac{\pi}{2}} (I_m \sin \omega t)^2 \cdot d(\omega t) \right]^{\frac{1}{2}}$$

$$= \frac{I_m}{2\sqrt{\pi}} \left[\left[\omega t - \frac{\sin 2\omega t}{2} \right]_0^{\frac{\pi}{2}} \right]^{\frac{1}{2}} = \frac{I_m}{2\sqrt{\pi}} \times \left(\frac{\pi}{2} \right)^{\frac{1}{2}}$$

$$\boxed{i_{T_1} = \frac{I_m}{2\sqrt{\pi}}}$$

$$\therefore \quad i_{T_1} = \frac{146.422}{2\sqrt{2}} = 51.776 \text{ Amp}$$

$$\therefore \quad \boxed{i_{T_1} = 51.776 \text{ Amp}}$$

r.m.s. value of diode current $+\left[\text{as diode current conducts for } \frac{\pi}{2} \text{ to } \pi \right]$

$$I_{D_1} = \frac{1}{2\pi} \int_0^{\frac{\pi}{2}} (I_m \sin \omega t)^2 \cdot d(\omega t)$$

$$I_{D_1} = \frac{I_m}{2\sqrt{2}}$$

$$\therefore \quad I_{D_1} = \frac{146.22}{2\sqrt{2}} = 51.776 \text{ Amp}$$

$$\therefore \quad \boxed{I_{D_1} = 51.776 \text{ Amp}}$$

Peak repetive voltage across thyristor = V_S = 230 V

Problem 6.4: A single phase full bridge inverter feeds power at 50 Hz to RLC load with $R = 5\ \Omega$, $L = 0.3$ H and $C = 50\ \mu$F. The d.c. input voltage is 220 V d.c.

(a) Find an expression for load current up to fifth harmonic. Also, calculate.

(b) Power absorbed by the load and the fundamental power.

(c) The r.m.s. and peak currents of each thyristor.
(d) Conduction time of thyristors and diodes if only fundamental component were considered.

Solution: (a) For $1 - \phi$ full bridge inverter output voltage is given as

$$V_o = \sum_{n=1,3,5}^{\infty} \frac{4V_S}{n\pi} \sin n\omega t \text{ volts}$$

or $\quad V_o = \dfrac{4V_S}{\pi} \sin\omega t + \dfrac{4V_S}{3\pi} \sin 3\omega t + \dfrac{4V_S}{5\pi} \sin 5\omega t$

$$= \frac{4V_S}{\pi} \left[\sin\omega t + \frac{\sin 3\omega t}{3} + \frac{\sin 5\omega t}{5} \right]$$

$$= \frac{4 \times 220}{\pi} \left[\sin\omega t + \frac{\sin 3\omega t}{3} + \frac{\sin 5\omega t}{5} \right]$$

$$= 280.11 \sin 314t + 93.37\sin (3 \times 314t)$$
$$+ 56 \sin (5 \times 314t)$$

Load impedance at frequency $n.f$ is

$$Z_n = 5 + J \left(2\pi \times 50 \times 0.3 \times n - \frac{10^6}{2\pi \times 50 \times 50n} \right)$$

$$\boxed{Z_n = 5 + J \left(94.24n - \frac{63.66}{n} \right)}$$

$\therefore \quad Z_1 = 5 + J (94.24 - 63.66)$
$\quad\quad = 5 + J\, 30.57 = 30.97 \angle 80.71°$

$\therefore \quad \boxed{Z_1 = 30.97 \angle 80.71°}$

Similarly, $\quad Z_3 = 5 + J (282.72 - 21.22)$
$\quad\quad\quad = 5 + J\, 261.5$

$$\boxed{Z_3 = 261.54 \angle 88.9°}$$

and $\quad Z_5 = 5 + J (471.2 - 12.732) = 5 + J\, 458.468$

$$\boxed{Z_5 = 458.49 \angle 89.37°}$$

∴ Load current is given by :

$$i_o = \frac{280.11}{30.97} \sin(\omega t - 80.71) + \frac{93.37}{261.54}$$

$$\sin(3\omega t - 88.9°) + \frac{56}{458.49} \sin(5\omega t - 89.37°)$$

∴
$$\boxed{\begin{aligned} i_o &= 9.04 \sin(\omega t - 80.71°) + 0.357 \sin(3\omega t - 88.9°) \\ &\quad + 0.122 \sin(5\omega t - 89.3) \end{aligned}}$$

(b) Peak load current

$$I_m = \sqrt{(9.04)^2 + (0.357)^2 + (0.122)^2}$$

$$\boxed{I_m = 9.0478 \text{ A}}$$

r.m.s. load current $= \dfrac{I_m}{\sqrt{2}} = \dfrac{9.0478}{\sqrt{2}}$

$$\boxed{I_{or} = 6.3978 \text{ A}}$$

∴ $\boxed{\text{Load power} = (I_{or})^2 \times 5 = 204.65 \text{ W}}$

r.m.s. value of fundamental load current

$$I_{o1} = \frac{I_{m_1}}{\sqrt{2}} = \frac{9.04}{\sqrt{2}} = 6.3922 \text{ A}$$

∴Fundamental load power

$$P_{o1} = I_{o1}^2 R = (6.3922)^2 \times 5$$

∴ $\boxed{P_{o1} = 204.304 \text{ W}}$

(c) Peak thyristor current $= I_m = 9.0478$ A

r.m.s. value of thyristor current $= \dfrac{I_m}{2} = 4.5239$ A

(d) Fundamental component of current is

$$i_{01} = 9.04 \sin(314\, t - 80.71°)$$

This current lags the fundamental voltage by 80.71°. This means diode conducts for 80.71° and thyristor for 180° − 80.71° = 99.29°.

∴ Conduction time for thyristor

$$= \frac{99.29° \times \pi}{180 \times 314} = 5.51 \times 10^{-3} = 5.51 \text{ m sec}$$

Conduction time for diode

$$= \frac{80.71 \times \pi}{180 \times 314} = 4.486 \times 10^{-3} = 4.486 \text{ m sec}$$

Hence conduction time for :

and
thyristor	= 5.51 m sec
diode	= 4.486 m sec

Problem 6.5: A single phase modified *MC* Murray invertor is fed by a d.c. source of 230 V. The d.c. source voltage may fluctuate by ± 25%. The load current during commutation may vary from 20 to 80 A. If thyristor turn-off time is 20 μ sec, calculate the values of *C* and *L*. Use a factor of safety of 2.

Also, obtain the value of resistance that gives critical damping.

Solution: The design is carried out on the basis of worst operating conditions which consist of minimum supply voltage V_{mn} and maximum load current I_{om}.

For modified half bridge MC Murray inverter it is found that normalized commutation energy $h(x)$ has minimum value of 0.446 when $x = 1.5$.

So, $\quad g(x) = 2 \cos^{-1}\left(\dfrac{1}{1.5}\right) = 1.682 = \dfrac{tc}{\sqrt{LC}}$ ...(1)

∴ $\quad V_{mn} \sqrt{\dfrac{C}{L}} = I_{CP} = x\, I_{om} = 1.5\, I_{om}$ ...(2)

or multiplying (1) and (2) gives.

$$C = \frac{1.5\, t_C\, I_{om}}{1.682\, V_{mn}} = 0.892\, \frac{t_C\, I_{om}}{V_{mn}}$$

∴ $\quad C = \dfrac{0.892 \times 20 \times 10^{-6} \times 80}{230\,(1 - 0.25)}$

or $\quad C = 8.27 \times 10^{-6} \text{F}$

or $\quad C = 8.27 \text{ μF}$

So, for full bridge MC Murray inverter capacitor required is double that of half bridge. So

$$C = 2 \times 8.27 \ \mu F$$

$$\boxed{C = 16.547 \ \mu F}$$

From the eq. (1) and (2) inductance is given as :

$$L = \frac{t_C \, V_{mn}}{1.682 \times 1.5 \, I_{om}} = 0.3964 \, \frac{t_C \, V_{mn}}{I_{om}}$$

$$\therefore \qquad L = \frac{0.3964 \times 20 \times 10^{-6} \times 230 \, (1 - 0.25)}{80}$$

$$L = 17.09475 \times 10^{-6} \ H$$

or $\qquad L = 17.09475 \ \mu H$

So, for full bridge MC Murray inverter Inductance is double with that of half bridge. So,

$$L = 2 \times 17.09475 \ \mu H$$

$$\boxed{L = 34.1895 \ \mu H}$$

The value of critical resistance is

$$R_d = 2 \sqrt{\frac{L}{C}} = 2 \times \sqrt{\frac{34.1895}{16.547}} = 2 \times 1.4374 \ \Omega$$

$$\boxed{R_d = 2.8748 \ \Omega}$$

Problem 6.6: For the series inverter control of voltage, two single-phase inverter are connected in series. Each inverter has output voltage of 400 V and each transformer has primary to secondary turns ratio of 1 : 2 calculate the resultant output voltage from this scheme in case firing angles for the two inverter differ by 30°.

Solution:

From the phaser diagram the output voltage is given as :

$$V_0 = \left[V_{o1}^2 + V_{o1}^2 + 2 \, V_{o1} \, V_{o2} \cos \theta \right]^{\frac{1}{2}}$$

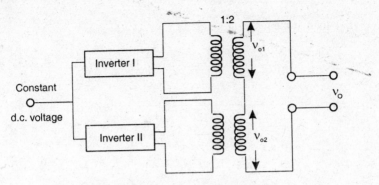

(a) Series inverter control of two inverter

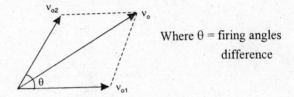

Where θ = firing angles difference

Fig. P. 6.6 (b). Phaser diagram of series inverter

Given output of the inverter I and II is 400 V

$\therefore$ $V_{o1} = 2 \times 400$ and $V_{o2} = 2 \times 400$

$\therefore$ $V_{o1} = V_{o2} = 800$ V

$\therefore$ $V_o = \left[(800)^2 + (800)^2 + 2 \times 800 \times 800 \cdot \cos 30^\circ \right]^{\frac{1}{2}}$

$\boxed{V_o = 1545.48 \text{ V}}$

Problem 6.7: For a single phase bridge inverter, source voltage is 230 V d.c. and the load is series RLC with $R = 1\ \Omega$, $\omega L = 2\ \Omega$ and $\dfrac{1}{\omega c} = 1.5\ \Omega$. The output voltage is controlled by single pulse modulation and the pulse width is 120°. Determine the magnitude of r.m.s. value of fundamental, third, fifth and seventh harmonics components of the output current.

Also find the power delivered to load.

Solution: r.m.s. value of 1 – ϕ full bridge inverter is given as :

$$V_0 = \sum_{n=1,3,5}^{\infty} \frac{4 V_S \sin n_d}{n\pi} \sin n\omega t \text{ volts}$$

load impedance at frequency $n.f$ is :

$$Z_n = 1 + J\left(2n - \frac{1.5}{n}\right)$$

$\therefore \qquad Z_i = 1 + J(2 - 1.5) = 1 + j\,0.5$

$$\boxed{Z_1 = 1.118 \ \angle\, 26.56^\circ}$$

$$Z_3 = 1 + J\left(6 - \frac{1.5}{6}\right)$$

$$\boxed{Z_3 = 5.83 \ \angle\, 80.134^\circ}$$

$$Z_5 = 1 + J\left(5 \times 2 - \frac{1.5}{5}\right)$$

$$\boxed{Z_5 = 9.75 \ \angle\, 84.11^\circ}$$

$$Z_7 = 1 + J\left(2 \times 7 - \frac{1.5}{7}\right)$$

$$\boxed{Z_7 = 13.82 \ \angle\, 85.85^\circ}$$

(1) r.m.s. value of fundamental load voltage

$$V_{o1} = \frac{4 V_S}{\pi.\sqrt{2}} \sin d = \frac{4 \times 230}{\pi\sqrt{2}} \sin 60^\circ \ [\therefore 2d = 120^\circ]$$

$$\boxed{V_{o1} = 179.33 \text{ V}}$$

$\therefore \qquad I_{o1} = \dfrac{V_{o1}}{Z_1} = \dfrac{179.33}{1.118} = 160.402 \text{ Amp}$

$\therefore \qquad \boxed{I_{o1} = 160.402 \text{ Amp}}$

(2) r.m.s. value of third harmonic voltage

$$V_{o3} = \frac{4 V_S}{\pi\sqrt{2}} \sin 3d = \frac{4 \times 230}{\pi\sqrt{2}} \sin 180^\circ$$

$$\boxed{V_{o3} \ = \ 0 \ \text{V}}$$

$\therefore$ $\boxed{I_{o3} \ = \ 0 \ \text{Amp}}$

(3) r.m.s. value of fifth harmonic voltage

$$V_{o5} \ = \ \frac{4V_S}{5\pi\sqrt{2}}\sin 5d \ = \ \frac{4\times 230}{5\pi\sqrt{2}}\sin 300 \ = \ -\frac{179.33}{5} \ \text{V}$$

$\therefore$ $I_{o5} \ = \ -\dfrac{179.33}{5\times 9.75} \ = \ -3.678 \ \text{Amp}$

$\therefore$ $\boxed{I_{o5} \ = \ -3.678 \ \text{Amp}}$

(4) r.m.s. value of seventh harmonic voltage

$$V_{o7} \ = \ \frac{4V_S}{7\pi\sqrt{2}}\sin 7\times 60°$$

$\boxed{V_{o7} \ = \ 25.618 \ \text{V}}$

$\therefore$ $I_{o7} \ = \ \dfrac{V_{07}}{Z_{07}} = \dfrac{25.618}{13.82}$

or $\boxed{I_{o7} \ = \ 1.853 \ \text{Amp}}$

So, r.m.s. value of total load current

$$= \ \sqrt{(160.402)^2 + (3.678)^2 + (1.853)^2} \ = \ 160.482$$

$\therefore$ Power delivered to load $= (160.482)^2 \times R = (160.482)^2 \times 1$

$$\boxed{P_o \ = \ 25754.7 \ \text{W}}$$

Problem 6.8: In a $1 - \phi$ *ASCI* with load L, SCR T_3, T_4 are conducting a constant current $I = 10$ A. If T_1, and T_2 are turned on at $t = 0$ to force commutate T_3, T_4. Find the time required for the load current to fall to zero. Load $L = 10$ μH and commutating capacitance $C = 6$ μF. Find also the total commutation interval and the circuit turn-off time for each of the SCR.

Solution: For $1-\phi$ ASCI with load L, the wave form of capacitor voltages $V_{o1} = V_{o2} = V_{CO}$ and load current is given as :

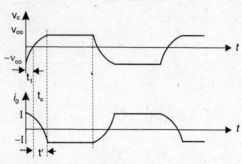

Fig. P. 6.8

From the given Fig. P. 6.8

The total commutation interval $t_C = t_1 + t_2$

or $\quad t_C = \dfrac{C}{2I} V_{CO} + \dfrac{\pi}{2\omega_o}$

$$\boxed{t_C = \left(1 + \dfrac{\pi}{2}\right) \sqrt{LC}}$$

$\therefore \quad t_C = \left(1 + \dfrac{\pi}{2}\right) \sqrt{10 \times 10^{-6} \times 6 \times 10^{-6}} = 2.57 \times \sqrt{60} \times 10^{-6}$

$\quad t_C = 19.913 \times 10^{-6}$

or $\quad \boxed{t_C = 10.913 \ \mu \ sec}$

Circuit turn-off time for each of the SCR is also equal to $t_1 = \dfrac{1}{\omega_o}$

$\therefore \qquad t_1 = \dfrac{1}{\omega_o} = \sqrt{LC}$

or $\qquad t_1 = \sqrt{LC}$

$\therefore \qquad t_1 = \sqrt{10 \times 10^{-6} \times 6 \times 10^{-6}} \ sec$

or $\qquad t_1 = \sqrt{60} \times 10^{-6} \ sec$

or $\qquad \boxed{t_1 = 7.746 \ \mu \ sec}$

since $\boxed{i_o = 2\,I\,\cos\,\omega_o t' - I}$

i_o is zero when $\cos\,\omega_o t' = \dfrac{1}{2}$

or $\quad\quad \omega_o t' = \cos^{-1}\,(0.5)$

$\quad\quad\quad\quad \omega_o t' = \dfrac{\pi}{3}$

or $\quad\quad \boxed{t' = \dfrac{\pi}{3\,\omega_o}}$

or $\quad\quad t' = \dfrac{\pi\sqrt{LC}}{3}$

$\quad\quad\quad\quad t' = \dfrac{\pi}{3}\,\sqrt{60}\,\times 10^{-6}\ \text{sec}$

or $\quad\quad \boxed{t' = 8.1115\ \mu\ \text{sec}}$

So, the total time required for the load current to fall to zero is equal to : $(t_1 + t')$

or $\quad\quad t_1 + t' = (7.746 + 8.1115)\ \mu\ \text{sec}$

$\boxed{t_1 + t' = 15.857\ \mu\ \text{sec}}$

Problem 6.9: A single phase auto-sequential commutated inverter is used to deliver power to load of $R = 12\ \Omega$ from a 240 V d.c. source. If the inverter output frequency is 60 Hz, thyristor turn-off time 15 μs and factor of safety 2, then determine the suitable values for source inductance and the commutating capacitance. Neglect all losses and assume a maximum current change of 0.4 in one cycle.

Solution: Time of one cycle,

$$T = \frac{1}{f} = \frac{1}{60}\ \text{sec}$$

$\therefore\quad\quad \boxed{T = \dfrac{1}{60}\ \text{sec}}$

$\therefore$ Rate of change of current is equal to

$$\frac{di}{dt} = \frac{0.4\,A}{T} = 0.4 \times 60 \text{ A/sec}$$

$$\boxed{\frac{di}{dt} = 24 \text{ A/sec}}$$

A short circuit at the load terminals of the inverter puts the most severe conditions on the source. So the value of source inductance must be obtained from these considerations :

$$\boxed{V_S = L \frac{di}{dt}}$$

or source inductance

$$L = \frac{V_S}{\dfrac{di}{dt}} = \frac{240\,V}{\dfrac{24\,A}{\sec}} = \frac{240\,V}{\dfrac{24\,A}{\sec}} = 10 \text{ H}$$

$$\therefore \qquad \boxed{L = 10 \text{ H}}$$

Since the circuit turn-off time is given as :

$$\boxed{t_C = RC \ln 2}$$

Where, 2 is the factor of safety.

Given, $t_C = (15 \ \mu \ C \times 2)$ and $R = 12 \ \Omega$

So, $2 \times 15 \times 10^{-6} = 12 \times C \ln 2$

or $\qquad C = \dfrac{2 \times 15 \times 10^{-6}}{12 \times \ln 2} \text{ F}$

or $\qquad C = \dfrac{2 \times 15 \times 10^{-6}}{8.3177} \text{ F}$

or $\qquad C = 1.803 \times 10^{-6} \text{ F} \times 2$

or $\qquad \boxed{C = 1.803 \ \mu F \times 2}$

or $\qquad C = 1.803 \times 2 \ \mu F$

or $\qquad \boxed{C = 3.606 \ \mu F}$

Problem 6.10: A single phase half-bridge inverter feeds a resistive load of $R = 7.2\ \Omega$. The d.c. voltage of the inverter $V_S = 200$ V. Determine:
(a) r.m.s. value of the fundamental component of the voltage at the output.
(b) Output power.
(c) Average and peak currents of the thyristor.
(d) Peak inverse voltage.

Solution: Given $V_S = 200$ V

The output voltage wave form is as shown in Fig. P. 6.10

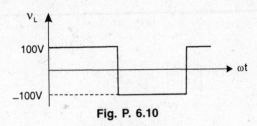

Fig. P. 6.10

Its fourier series can be expressed as

$$V_o = \sum_{n=1,3,5}^{\infty} \frac{2V_S}{n\,\pi} \sin n\omega t$$

(a) r.m.s. value of fundamental component of voltage is

$$= \frac{2 \times 100}{1 \times \pi} \times \frac{1}{\sqrt{2}} = 45 \text{ V}$$

(b) Output power $= \dfrac{V_S^2}{R} = \dfrac{(100)^2}{4 \times 7.2} = 347.2$ W

(c) Peak value of thyristor current $= \dfrac{50}{7.2} = 6.94$ A

and average of thyristor current $= 0.5 \times 6.94 = 3.472$ Amp
(d) Peak inverse voltage of the thyristor is $= 100$ V

Problem 6.11: A single phase full-bridge inverter has RLC load of $R = 5\ \Omega$, $L = 30$ mH and $C = 150\ \mu$F. The d.c. input voltage is 230 V and the output frequency is 50 Hz.
(a) Find an expression for load current up to fifth harmonic. Also calculate.
(b) r.m.s. value of fundamental load current.

(c) The power absorbed by load and the fundamental power.
(d) The r.m.s. and peak currents of each thyristor.
(e) Conduction time of thyristors and diodes if only fundamental component were considered.

Solution: (a) For a single phase full-bridge inverter output voltage is expressed as

$$V_o = \sum_{n=1,3,5}^{\infty} \frac{4V_S}{n\pi} \sin n\omega t \text{ volts}$$

For upto 5th harmonics

$$V_o = \frac{4V_S}{\pi} \sin \omega t + \frac{4V_S}{3\pi} \sin 3\omega t + \frac{4V_S}{5\pi} \sin 5\omega t$$

$$V_o = \frac{4 \times 230}{\pi} \left[\sin \omega t + \frac{1}{3} \sin 3\omega t + \frac{1}{5} \sin 5\omega t \right]$$

$$= 292.85 \sin 314t + 97.62 \sin (942t) + 58.57 \sin 1570t$$

Load impedance at frequency $n.f$ is

$$Z_n = 5 + J \left(2\pi \times 50 \times 30 \times 10^{-3} n - \frac{10^{-6}}{2\pi \times 50 \times 150n} \right)$$

$$= 5 + J \left(9.42n - \frac{21.22}{n} \right)$$

$$\therefore \quad Z_1 = \sqrt{5^2 + (9.42 - 21.22)^2}$$

$$Z_1 = 12.81 \ \Omega$$

and

$$\phi_1 = \tan^{-1} \left(\frac{9.42 - 21.22}{5} \right)$$

$$\phi_1 = -67^\circ$$

Similarly,

$$Z_3 = \sqrt{5^2 + \left(9.42 \times 3 - \frac{21.22}{3} \right)^2}$$

$$Z_3 = 21.76 \ \Omega$$

$$\phi_3 = \tan^{-1} \frac{\left(3 \times 9.42 - \frac{21.22}{3} \right)}{5}$$

$$\phi_3 = 76.72°$$

Similarly, $Z_5 = \sqrt{5^2 + \left(9.42 \times 5 - \dfrac{21.22}{5}\right)^2}$

$$Z_3 = 43.15 \ \Omega$$
$$\phi_3 = 83.34°$$

∴ load current from the above equation is given by

$$i_o = \frac{292.85}{12.81} \sin(\omega t + 67) + \frac{97.62}{21.76} \sin(\omega t - 76.72°) +$$

$$\frac{58.57}{43.15} \sin(\omega t - 83.34°)$$

$$\boxed{i_o = 22.86 \sin(\omega t + 67°) + 4.48 \sin(\omega t - 76.72°) + 1.35 \sin(\omega t - 83.34°)}$$

(b) r.m.s. value of fundamental load current

$$I_{o1} = \frac{I_{m1}}{\sqrt{2}} = \frac{22.86}{\sqrt{2}} = 16.164 \ A$$

(c) Peak load current

$$I_m = \sqrt{(22.86)^2 + (4.48)^2 + (135)^2}$$

$$\boxed{I_m = 23.33 \ A}$$

r.m.s. load current $= \dfrac{23.33}{\sqrt{2}} = 16.49$ Amp

load power $= (16.49)^2 \times 5 = 1361.18$ W

(d) Peak thyristor current

$$\boxed{I_m = 23.33 \ \text{Amp}}$$

r.m.s. value of thyristor current

$$= \frac{23.33}{2} = 11.665 \ \text{Amp}$$

(e) Fundamental component of current is

$$\boxed{i_{o1} = 22.86 \sin(314t + 67°)}$$

This current leads the fundamental voltage component by 67°. This means that diode conducts for 67° and thyristor for 180° − 67° = 113°.

$$\therefore \text{Conduction time for thyristor} = \frac{113 \times \pi}{180 \times 314} = 6.27 \text{ ms}$$

$$\text{and conduction time for diodes} = \frac{67 \times \pi}{180 \times 314} = 3.73 \text{ ms}$$

In case SCR turn-off time is less than 3.73 ms, load commutation will occur and no forced commutation will be required under the assumption of no harmonics.

Problem 6.12: In a self commutated SCR circuit, the load consists of $R = 10\ \Omega$ in series with commutating components of $L = 10$ mH and $C = 10\ \mu F$. Check whether the circuit will commutate by itself when triggered from zero voltage condition on the capacitor. What will be the voltage across the capacitor and inductor at the time of commutation?

Find also $\left. \dfrac{di}{dt} \right|_t = 0$

Solution: Given $R = 10\ \Omega$, $L = 10$ mH and $C = 10\ \mu F$.

$$R^2 = (10)^2 = 100 \text{ and } \frac{4L}{C} = \frac{4 \times 10 \times 10^{-3}}{10 \times 10^{-6}} = 4000$$

As $R^2 < \dfrac{4L}{C}$, the circuit is under damped. So the series circuit will commutate on its own when triggered from zero voltage condition on the capacitor.

$$\text{Also } \xi = \frac{R}{2L} = \frac{10 \times 1000}{2 \times 10} = 500$$

$$\omega_o = \frac{1}{\sqrt{LC}} = \frac{1}{\sqrt{10 \times 10^{-3} \times 10 \times 10^{-6}}} = 3.162 \times 10^3 \text{ rad/sec}$$

$$\omega_r = \sqrt{\omega_o^2 - \xi^2} = [10^7 - 2.5 \times 10^7]^{\frac{1}{2}}$$

$$= 3.122 \times 10^3 \text{ rad/sec}$$

$$\psi = \tan^{-1}\left(\frac{\xi}{\omega_r}\right) \ \tan^{-1}\left(\frac{500}{3.122 \times 10^3}\right)$$

$$\boxed{\psi = 9.09^\circ}$$

Since the load current is zero, i.e. SCR will commutate when $\omega_r t = \pi$

or
$$t = \frac{\pi}{\omega_r} = \frac{\pi}{3.122 \times 10^3} = 1.006 \text{ ms}$$

$$\xi t = 500 \times 1.006 \times 10^{-3} = 0.503$$

The voltage across L at the time of commutation is

$$V_L = V_S \ \frac{\omega_o}{\omega_r} \ e^{-\xi t} \ \cos(\omega_r t + \psi)$$

$$= V_S \cdot \frac{3.162}{3.122} \ e^{-0.503} \ \cos(180^\circ + 9.097^\circ)$$

$$\boxed{V_L = -0.604 \ V_S}$$

The voltage across capacitor C at the time of commutation is

$$V_C = V_S \left[1 - e^{-\xi t} \frac{\omega_o}{\omega_r} \cos(\omega_r t - \psi) \right]$$

$$= V_S \left[1 - e^{-0.503} \times \frac{3.162}{3.122} \cos(180^\circ - 9.097^\circ) \right]$$

$$\boxed{V_C = 1.604 \ V_S}$$

Since,
$$\frac{di}{dt} = \frac{V_S}{\omega_r L} \left[e^{-\xi t} \omega_r \cos\omega_r t - \xi e^{-\xi t} \sin\omega_r t \right]$$

$$\left.\frac{di}{dt}\right|_{t=0} = \frac{V_S}{\omega_r L} \left[1 \cdot \omega_r \cdot 1 - 0 \right]$$

$$\left.\frac{di}{dt}\right|_{t=0} = \frac{V \omega_r}{\omega_r L} = \frac{V_L}{L}$$

or
$$\boxed{\left.\frac{di}{dt}\right|_{t=0} = \frac{V_S}{10 \times 10^{-3}} = 100 \ V_S \ A/S}$$

Problem 6.13: Calculate the output frequency of a series inverter with the following parameters. Inductance $L = 6$ mH, capacitance $C = 1.2$ μF, Load resistance $R = 100$ Ω, $T_{off} = 0.2$ ms.

If the load resistance is varied from 40 to 140 Ω, find out the range of output frequency.

Solution: The time period of oscillation is given as

$$T' = \frac{\pi}{\sqrt{\dfrac{1}{LC} - \left(\dfrac{R}{2L}\right)^2}} = \frac{\pi}{\sqrt{\dfrac{1}{6 \times 1.2 \times 10^{-9}} - \dfrac{100^2}{4 \times 36 \times 10^{-6}}}} = 0.377 \text{ ms.}$$

The output frequency is given as

$$f = \frac{1}{2(T' + T_{off})} = \frac{1}{2(0.377 + 0.2) \times 10^{-3}} = 866.55 \text{ Hz}$$

When, $R = 40$ Ω, output frequencey.

$$= \frac{1}{\dfrac{2\pi}{\left[\dfrac{10^9}{6 \times 1.2} - \dfrac{1600 \times 10^6}{4 \times 36}\right]^{\frac{1}{2}}} + 0.4 \times 10^{-3}} = 1046.2 \text{ Hz}$$

Similarly, when $R = 140$ Ω, output frequency.

$$= \frac{1}{\dfrac{2\pi}{\left[\dfrac{10^9}{6 \times 1.2} - \dfrac{140^2 \times 10^6}{4 \times 36}\right]^{\frac{1}{2}}} + 0.4 \times 10^{-3}} = 239.8 \text{ Hz}$$

∴ Range of output frequency varies as : $\boxed{239.8 \text{ Hz to } 1046.2 \text{ Hz}}$

Problem 6.14: In a single-phase capacitor commutated parallel inverter using two thyristor and a centre tapped transformer, the source voltage is 220 V d.c. The centre tapped transformer has a turns ratio from each half primary winding to secondary winding of 2 : 1. For a load resistance of 10 Ω, find the value of capacitor C to obtain

20 μs turn-off time on the thyristor. Assume the inductor L large and transformer ideal. Take factor of safety as 2.

Solution: The voltage across capacitor C is given by

$$V_C = 2\,V_S\left[2\exp\left(-\frac{n^2 t}{4\,RC}\right) - 1\right]$$

Here, $n = \dfrac{1}{2}$, circuit turn-off time

$$t_C = 2 \times 20 = 40\ \mu s$$

The circuit turn-off time is obtained when V_C reduces to zero from $2\,V_S$. This gives

$$C = \frac{n^2 t_C}{4\,R\ln 2} = \frac{\left(\dfrac{1}{2}\right)^2 \times 40 \times 10^{-6}}{4 \times 10\ln 2} = 0.36 \times 10^{-6}$$

or $\boxed{C = 0.36\ \mu F}$

CHAPTER 7

ELECTRIC DRIVES

An electric motor together with its control equipment and energy transmitting device is called an electric drive. An electric drive together with its working machine constitutes an electric-drive system.

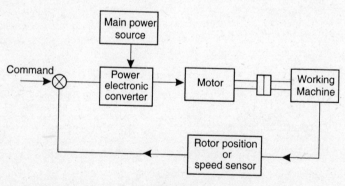

Fig. 7.1

Block diagram for a modern electric drive system.

Electric drives are mainly of two types: (1) d.c. drives, (2) a.c. drives.

d.c. drives

d.c. motor are widely used in adjustable speed drives and position control applications. Their speeds below base speed can be controlled by armature-voltage control and the speeds above base speed can be controlled by field flux control method. As speed controls methods for d.c. motor are simpler and less expansive, so d.c. motors are preterred where wide-speed control rage is required as compared to a.c. motors.

Phase controlled converters provide an adjustable d.c. output voltage from a fixed a.c. input voltage. d.c. chopper also provides d.c. output voltage from a fixed d.c. input voltage. The d.c. drives can be classified as :

(1) $1 - \phi$ d.c. drives
(2) $3 - \phi$ d.c. drives
(3) Chopper drives

Basic performance of d.c. motors

(a) Separately - excited d.c. motor

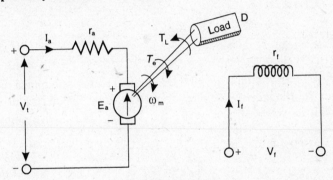

Fig. 7.2

D = Viscous friction constant (Nm-sec/rad)

For field circuit :
$$V_f = I_f \cdot r_f$$

For armature circuit

or
$$\boxed{V_t = E_a + I_a r_a}$$

$$E_a = K_a \phi \, \omega_m = K_m \, \omega_m$$
$$T_e = K_a \phi \, I_a = K_m \, I_a$$

$$\boxed{T_e = D\omega_m + T_L}$$

Angular speed; $\omega_m = \dfrac{V_t - I_a r_a}{K_m}$

or
$$\boxed{\omega_m = \frac{V_t - I_a r_a}{K_a \phi}}$$

speed is controlled by varying (1) armature terminal voltage (V_t) or by (2) field flux (ϕ).

(b) d.c. series motor

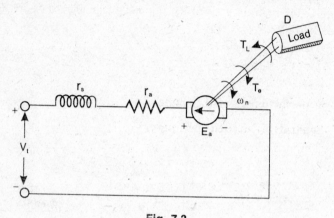

Fig. 7.3

$$V_t = E_a + I_a (r_a + r_s)$$
$$T_e = K_a \phi I_a$$

For no saturation, $\phi = C I_a$

$\therefore$ $$T_e = K_a C I_a^2 = K I_a^2$$

$$\boxed{\omega_m = \frac{V_t - I_a (r_a + r_S)}{K I_a}}$$

Here $K = K_{ac}$ = a constant in Nm/A^2 or in V-S/A rad.

For speed control below base speed, armature terminal voltage V_t is varied with I_a kept const. For speed control above base speed, series field flux is decreases by the use of diverter or tapped field control and I_a is kept constant.

(c) d.c. shunt motor

In d.c. shunt motor speed is generally constant and drops very little as load increases. This method is not widely used for speed control.

1 – ϕ d.c. drives : Depending upon the type of power electronic converter used in the armature circuit, 1 – ϕ d.c. drives may be classified as :

(a) 1 – ϕ half-wave converter drives.

(b) $1 - \phi$ semi converter drives.

(c) $1 - \phi$ full converter drives.

(d) $1 - \phi$ dual converter drives.

(a) $1 - \phi$ *half-wave converter drives*

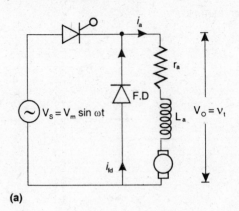

(a)

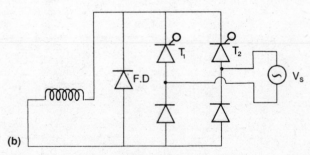

(b)

Fig. 7.4 (a) and (b)

Output voltage of converter V_o = armature terminal voltage V_t is given by

$$V_o = V_t = \frac{V_m}{2\pi} (1 + \cos \alpha_1)$$

Where, V_m = maximum source voltage

α_1 = firing angle of armature converter.

r.m.s. value of armature current $I_{ar} = I_a$

r.m.s. value of source or thyristor current

$$I_{sr} = \sqrt{I_a^2 \frac{(\pi - d_1)}{2\pi}} = I_a \sqrt{\frac{(\pi - d_1)}{2\pi}}$$

r.m.s. value of free wheel diode current is

$$I_{fdr} = I_a \sqrt{\frac{(\pi + \alpha_1)}{2\pi}}$$

Power factor $\boxed{P \cdot f = \dfrac{V_t \cdot I_a}{V_S \cdot I_{Sr}}}$

(b) $1 - \phi$ *semi converter drives*

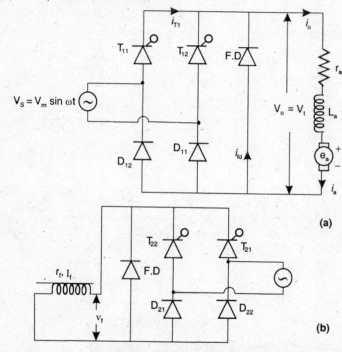

Fig. 7.5 (a) and (b)

$$V_o = V_t = \frac{V_m}{\pi} (1 + \cos \alpha_1)$$

$$V_f = \frac{V_m}{\pi} (1 + \cos \alpha_2)$$

r.m.s. value of source current

$$I_{sr} = I_a \left[\frac{\pi - \alpha}{\pi} \right]^{\frac{1}{2}}$$

r.m.s. value of free wheeling diode current

$$I_{fdr} = I_a \left[\frac{\alpha}{\pi} \right]^{\frac{1}{2}}$$

Input power factor

$$P \cdot f = \frac{V_t \cdot I_a}{V_S \cdot I_{Sr}}$$

This $1 - \phi$ semiconverter is also called $1 - \phi$ half-controlled bridge converter.

(c) $1 - \phi$ *full converter drives*

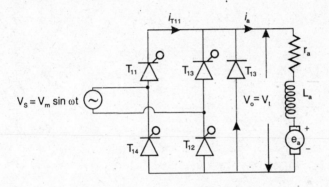

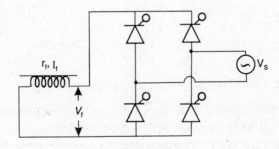

Fig. 7.6

$1 - \phi$ full converter is a two-quadrant drive. It is generally used for regenerative braking of the motor because it is possible to feed back power from motor to the a.c. source, which is necessary for regenerative braking.

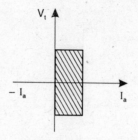

Fig. 7.7

Average output voltage for :

Armature converter : $\boxed{V_o = V_t = \dfrac{2\,V_m}{\pi}\,\cos\,\alpha_1}$ for $0 < \alpha_1 < \pi$

For the field converter, $\boxed{V_f = \dfrac{2\,V_m}{\pi}\,\cos\,\alpha_2}$ for $0 < \alpha_2 < \pi$

r.m.s. value of source current

$$\boxed{I_{sr} = I_a}$$

r.m.s. value of thyristor current

$$\boxed{I_{tr} = \frac{I_a}{\sqrt{2}}}$$

Input power factor $= \dfrac{V_t \cdot I_a}{V_S \cdot I_{Sr}} = \dfrac{2\sqrt{2}}{\pi}\,\cos\,\alpha_1$

(d) $1 - \phi$ *dual converter drives*

A $1 - \phi$ dual converter is obtained by connecting two full- converters in anti-parallel. It offers four-quadrant operation.

For converter $\perp$ in operation,

$$V_t = \frac{2\,V_m}{\pi}\cos\,\alpha_1 \text{ for } 0 \le \alpha_1 \le \pi$$

For converter 2 in operation,

$$V_t = \frac{2\,V_m}{\pi}\cos\,\alpha_2 \text{ for } 0 \le \alpha_2 \le \pi$$

Where, $\alpha_1 + \alpha_2 = \pi$

(1) Converter 1 with $\alpha_1 < 90°$, operates the motor in forward motoring mode in quadrant 1.

(2) Converter 1 with $\alpha_1 > 90°$, and with field excitation reversed operates the motor in forward regenerative braking mode in quadrant 4.

(3) Converter 2 with $\alpha_2 < 90°$, operates the motor in reverse motoring mode in quadrant 3.

(4) Converter 2 with $\alpha_2 > 90°$ and with field excitation reversed operates the motor in reverse regenerative braking mode in quadrant 2.

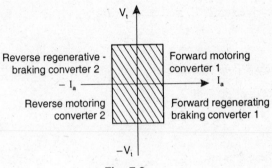

Fig. 7.8

3 – φ d.c. drives

1 – φ d.c. drives are limited upto 15 kW. For high power 3 – φ d.c. drives are used. As 1 – φ d.c. drives, 3 – φ d.c. drives are also classified into four types.

1. 3 – φ half wave converter drives
2. 3 – φ semi converter drives
3. 3 – φ full converter drives
4. 3 – φ dual converter drives

Average output voltage in

1. 3 – φ *half wave converter drives*

$$V_o = V_t = \frac{3 V_{ml}}{2 \pi} \cos \alpha_1, \text{ for } 0 \le \alpha_1 \le \pi$$

2. 3 – φ *semi converter drives*

For converter $\perp$, $V_o = V_t = \dfrac{3V_{ml}}{2\pi} (1 + \cos \alpha_1)$

$$\text{for } 0 < \alpha_1 < \pi$$

For converter 2, $V_f = \dfrac{3V_{ml}}{2\pi} (1 + \cos \alpha_2)$ for $0 < \alpha_2 < \pi$

3. 3 – ϕ *full converter drives*

For converter 1, $V_o = V_t = \dfrac{3V_{ml}}{\pi} \cos \alpha_1$ for $0 \leq \alpha_1 \leq \pi$

For converter 2, $V_f = \dfrac{3V_{ml}}{\pi} \cos \alpha_2$ for $0 \leq \alpha_2 \leq \pi$

4. 3 – ϕ *dual converter drives*

For converter 1 or 2

$$V_o = V_t = \dfrac{3V_{ml}}{\pi} \cos \alpha_f \qquad\qquad \text{for } 0 \leq \alpha_f \leq \pi$$

For field circuit

$$V_f = \dfrac{3V_{ml}}{\pi} \cos \alpha_f \qquad\qquad \text{for } 0 \leq \alpha_f \leq \pi$$

Chopper drives

When variable d.c. voltage is to be obtained from fixed d.c. voltage, d.c. choppers are used chopper is easily adaptable for regenerative braking of d.c. motors and thus kinetic energy of the drive can be returned to the d.c. source. This results in overall energy saving which is required in transportation systems for frequent stops. So the use of d.c. chopper in traction system is now accepted all over the world. Chopper drives are also used in battery-operated vehicles, dynamic braking and for combination of regenerative and dynamic control of d.c. drives. There are two chopper control methods :

(1) Power control or motoring control.
(2) Regenerative-braking control.

(1) Power control or motoring control

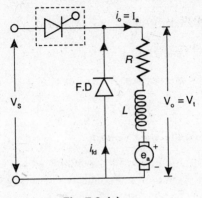

Fig 7.9 (a)

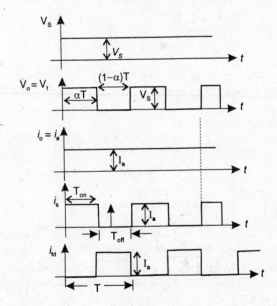

Fig. 7.9 (b)

Average motor voltage

$$V_o = V_t = \frac{T_{on}}{T} \cdot V_S = \alpha V_S = f T_{on} \cdot V_S$$

Where, $\alpha = \dfrac{T_{on}}{T}$ = duty cycle

$$f = \text{chopping frequency} = \frac{1}{T}$$

Power delivered to motor $= V_t I_a = \alpha\, V_S\, I_a$

Average source current $= \dfrac{T_{on}}{T}\, I_a = \alpha\, I_a$

Input power to chopper $=$ (average input voltage) $\times$
(ave. source current)
$= V_S \cdot \alpha I_a$

(2) *Regenerative-breaking control*

In regenerative-braking control, the motor acts as a generator and the kinetic energy of the motor and connected load is returned to the supply.

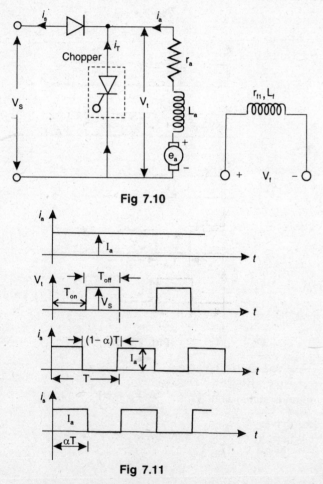

Fig 7.10

Fig 7.11

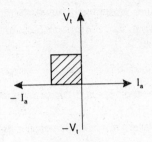

Fig 7.12

Average voltage across chopper is

$$V_t = \frac{T_{\text{off}}}{T} \cdot V_S = (1 - \alpha) \, V_S$$

Power generated by the motor $= V_t \cdot I_a = (1 - \alpha) \, V_S \cdot I_a$

Motor emf generated

$$E_a = K_m \, \omega_m = V_t + I_a r_a = (1 - \alpha) \, V_S + I_a r_a$$

Motor speed during regenerative braking

$$\boxed{\omega_m = \frac{(1 - \alpha) \, V_S + I_a r_a}{K_m}}$$

Minimum braking speed

$$\omega_{mn} = \frac{I_a r_a}{K_m}$$

Maximum braking speed

$$\omega_{mx} = \frac{V_S + I_a r_a}{K_m}$$

Thus regenerative braking control is effective only when motor speed is less than ω_{mx} and greater than ω_{mn}.

$$\therefore \quad \boxed{\omega_{mn} < \omega_m < \omega_{mx}}$$

Note : Regenerative braking of chopper-fed separately excited d.c. motor is stable and for d.c. series motor is unstable.

a.c. drives

Advantages of a.c. drives over d.c. drives :

(1) Low maintenance is requied in a.c. drives.

(2) a.c. motors are less expansive as compare to d.c. drives.

(3) For same rating, a.c. motors are lighter in weight as compared to d.c. motors.

Disadvantage of a.c. drives over d.c. drives :

(1) Power converters for a.c. drives generate harmonics in the circuit.

(2) Power converters for a.c. drives are more expansive.

In general there are two types of a.c. drives :

(1) Induction motor drives.

(2) Synchronous motor drives.

Induction motor drives

The various methods of speed control of $3 - \phi$ induction motor through semiconductor devices are :

(1) Stator voltage control

(2) Stator frequency control

(3) Stator voltage and frequency control

(4) Stator current control

(5) Static rotor-resistance control

(6) Slip-energy recovery control.

Methods (1) to (4) are applicable for both squirrel cage I.M and wound rotor I.M where as (5) and (6) can be used for wound rotor I.M only.

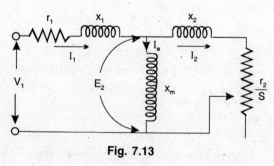

Fig. 7.13

$$I_2 = \frac{V_1}{\left(\dot{r_1} + \dfrac{r_2}{s}\right) + J\left(x_1 + x_2\right)}$$

air-gap power

$$P_g = 3 I_2^2 \frac{r_2}{s}$$

Mechanical power developed $P_m = (1 - S) P_g$

Developed torque in rotor

$$T_e = \frac{P_m}{\omega_m} = \frac{3}{\omega_s} \frac{V_1^2}{\left(r_1 + \frac{r_2}{s}\right)^2 + (x_1 + x_2)^2} \cdot \frac{r_2}{s}$$

Output or shaft power

$$P_{sh} = P_m - \text{fixed loss (friction and winding loss)}$$

$\therefore \qquad T_{sh} = \dfrac{P_{sh}}{\omega_m} = \dfrac{P_{sh}}{\omega_m(1-s)}$

Slip at which maximum torque occurs is given by

$$S_m = \frac{r_2}{\sqrt{r_1^2 + (x_1 + x_2)^2}}$$

$\therefore$ Maximum torque

$$T_{e.m} = \frac{3}{2\,\omega_s} \left[\frac{V_1^2}{r_1 + \sqrt{r_1^2 + (x_1 + x_2)^2}} \right]$$

$$\frac{T_e}{T_{e.m}} = \frac{2}{\dfrac{s}{s_m} + \dfrac{s_m}{s}}$$

If, $S < S_m$, then
$$\boxed{\frac{T_e}{T_{e.m}} = \frac{2 \cdot s}{s_m}} \qquad \qquad \text{...(1)}$$

Motor speed $\omega_m = \omega_S \left[1 - \dfrac{S_m \cdot T_e}{2 \, T_{em}} \right]$ \qquad \text{...(2)}

Form eq. (1) and eq. (2) we get

$$\boxed{\omega_m = \omega_S \left[1 - \frac{r_2}{x_1 + x_2} \frac{T_e}{2 \, T_{em}} \right]}$$

This expression shows that drop in speed from no load to full load depends on rotor resistance.

Synchronous motor drives

A synchronous motor is a constant-speed machine and always rotates with zero slip at the synchrous speed $\left(N_S = \dfrac{120f}{P} \right)$, which depends on the frequency and the number of poles. The power factor of a synchronous machine can be controlled by varying the field current. The cycloconverters and inverters are widening the application of synchronous motors in variable-speed drives. The synchronous motor can be classified into six types.

(1) Salient-pole motors
(2) Cylindrical-rotor motors
(3) Reluctance motors
(4) Permanent-magnet motors
(5) Switched reluctance motor
(6) Brushless d.c. and a.c. motors

Voltage/frequency control of synchrous motors

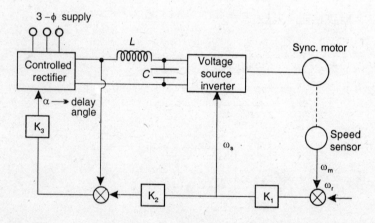

Fig 7.14

V/f control of synchronous motor.

PROBLEMS

Problem 7.1: The speed of a separately excited d.c. motor is controlled through $1 - \phi$ half wave controlled converter from 230 V mains. The motor armature resistance is 0.5 Ω and motor constant is $K = 0.4$ V.s/rad. For load torque of 20 Nm at 1500 r.p.m. and for constant armature current, calculate (i) firing angle delay of the converter (ii) r.m.s value of thyristor current and (iii) input $P.f$ of the motor.

Solution: Armature current $I_a = \dfrac{T_e}{K_m} = \dfrac{20}{0.4} = 50$ Amp

$\therefore$
$$\boxed{I_a = 50 \text{ Amp}}$$

Motor e.m.f. $E_a = K_m \, \omega_m = 0.4 \times 2\pi \times \dfrac{1500}{60}$

$$\boxed{E_a = 62.83 \text{ V}}$$

(i) For $1 - \phi$ half-wave controlled converter

$$V_t = \frac{V_m}{2\pi} (1 + \cos \alpha_1) = E_a + I_a r_a$$

$$= \frac{\sqrt{2} \times 230}{2\pi} (1 + \cos \alpha_1) = 62.83 + 20 \times 0.5$$

$$= \frac{\sqrt{2} \times 230}{2\pi} (1 + \cos \alpha_1) = 87.83$$

or $\quad (1 + \cos \alpha_1) = 1.696$

or $\quad\quad \cos \alpha_1 = 0.696$

or $\quad\quad\quad \alpha_1 = \cos^{-1}(0.696)$

$$\boxed{\alpha_1 = 45.84^\circ}$$

(b) r.m.s. value of thyristor current is

$$I_{Tr} = I_a \left(\frac{\pi - \alpha_1}{2\pi} \right)^{\frac{1}{2}} = 50 \left(\frac{180^\circ - 45.84^\circ}{360^\circ} \right)^{\frac{1}{2}}$$

$$= 50 \, (0.372)^{\frac{1}{2}} = 50 \times 0.610$$

$$\boxed{I_{Tr} = 30.52 \text{ Amp}}$$

(c) Input power factor of armature converter is given as :

$$(P.f)_{input} = \frac{V_t \cdot I_a}{V_S \cdot I_{Tr}} = \frac{87.83 \times 50}{230 \times 30.52}$$

$$\boxed{(P.f)_{input} = 0.6256}$$

Problem 7.2: A separately excited d.c. motor, operating from a $1-\phi$ half controlled bridge at a speed of 1500 r.p.m. has an input voltage of $300 \sin 314t$ and a back e.m.f. 80 V. The SCRs are fired symmetrically at $\alpha = 30°$ in every half cycle and the armature has a resistance of 5 Ω. Calculate the average armature current and the motor torque.

Solution: For a $1-\phi$ semiconverted feeding to a separately excited d.c. motor is given as :

$$V_o = V_t = \frac{V_m}{\pi} (1 + \cos \alpha_2) = E_a + I_a r_a$$

or $\qquad \dfrac{300}{\pi} (1 + \cos 30°) = 80 + I_a \times 5$

or $\qquad\qquad 178.19 = 80 + I_a \times 5$

or $\qquad\qquad 98.19 = 5\, I_a$

or $\qquad\qquad\boxed{I_a = 19.63 \text{ Amp}}$

Now, the motor e.m.f. is given as :

$$E_a = K_m \omega_m = K_m \times \frac{2\pi \times 1500}{60}$$

or $\qquad 80 = K_m \times \dfrac{2\pi \times 1500}{60}$

or $\qquad \boxed{K_m = 0.5 \text{ V-S/rad}}$

$\therefore$ Motor torque

$$T_e = K_m I_a = 0.5 \times 19.63$$

$$\boxed{T_e = 9.99 \text{ Nm}}$$

Problem 7.3: A separately excited d.c. motor has its armature circuit connected to one semiconverter and field winding to another semiconverter. The supply for both the converters is $1 - \phi$ 230 V, 50 Hz. Resistance for the field circuit is 100 Ω and that for the armature circuit is 0.2 Ω. Rated load torque is 80 Nm at 1000 r.p.m. The motor constant is 0.8 V-s/A rad and magnetic saturation is neglected. For ripple free armature and field currents and with zero degree firing angle for field converter, determine (a) rated armature current, (b) firing-angle delay of armature converter at rated load, (c) speed regulation at full load, (d) input p.f of the armature converter and the drive at rated load.

Solution: Field output voltage V_f at $\alpha_2 = 0°$ is

$$V_f = \frac{2V_m}{\pi} = \frac{2 \times \sqrt{2} \times 230}{\pi}$$

or $\boxed{V_f = 207.8 \text{ V}}$

$\therefore \qquad I_f = \dfrac{207.8}{100} = 2.078 \text{ Amp}$

(a).$\therefore \qquad T_e = 80 \text{ Nm} = (KI_f) \cdot I_a$

$\qquad\qquad 80 = (0.8 \times 2.078) \cdot I_a$

or $\qquad I_a = \dfrac{80}{1.656}$

or $\qquad \boxed{I_a = 48.31 \text{ Amp}}$

(b) Armature output voltage is given as :

$$V_t = (KI_f) \cdot \omega_s + I_a \cdot r_a$$

i.e. $\qquad \boxed{V_t = E_b + I_a r_a}$

or $\qquad V_t = (0.8 \times 2.07) \cdot 2\pi \dfrac{1000}{60} + 48.31 \times 0.2$

$\qquad\qquad = (173.41 + 9.662) \text{ V}$

$\boxed{V_t = 183.07 \text{ V}}$

$\therefore \qquad V_t = \dfrac{V_m}{\pi} (1 + \cos \alpha_1) = 183.07 \text{ V}$

or $\dfrac{\sqrt{2} \times 230}{\pi}(1 + \cos \alpha_1) = 183.07$ V

or $(1 + \cos \alpha_1) = 1.768$

or $\boxed{\alpha_1 = 39.80^\circ}$

(c) At no load $I_a = 0$ and let speed be N_o.

So, $V_t = 183.07 = (KI_f) \cdot \dfrac{2\pi N_o}{60}$

or $183.07 = (0.8 \times 2.07) \cdot \dfrac{2\pi N_o}{60}$

or $\boxed{N_o = 1055.69 \text{ r.p.m.}}$

So, Speed regulation is given as :

$$= \dfrac{N_o - N_{\text{full}}}{N_{\text{full}}} \times 100 \% = \dfrac{1055.69 - 1000}{1000} \times 100 \%$$

$$\boxed{\text{Speed regulation} = 5.569\%}$$

(d) Input *p.f* of the armature converter

$$\dfrac{V_t \cdot I_a}{V_S \cdot I_{ar}} = \dfrac{183.07 \times 48.31}{230 \times 48.31}$$

$$\boxed{p.f = 0.796}$$

Total r.m.s. current taken from the source

$$I_{sr} = \sqrt{(I_{ar})^2 + (I_{fr})^2} = \sqrt{(48.31)^2 + (2.07)^2}$$

$$\boxed{I_{sr} = 48.35 \text{ Amp}}$$

$\therefore$ $V_A = V_S \times I_{sr} = 230 \times 48.35$ $V_A = 11120.5$ V_A

with no loss in the converter, total power input to motor and field

$$= V_t \cdot I_a + V_f \cdot I_f = 183.07 \times 48.31 + 207.8 \times 2.078$$

$$\boxed{\text{Power input} = 9274.25 \text{ watts}}$$

. Input *p.f* of the drive $= \dfrac{\text{Power input in Watt}}{\text{Input in } V_A} = \dfrac{9274.25}{11120.5} = 0.8335$ lag

Problem 7.4: A separately excited d.c. motor is fed from two 1 – ϕ semiconverters, one in the armature circuit and the other in the field circuit. Field current is constant at 2 A. Motor armature resistance is 0.8 Ω and motor constant is $K = 0.5$ V-s/A rad. a.c. voltage is 230 V, 50 Hz, for a ripple-free armature current and speed of 1500 r.p.m. calculate:

(a) Motor current and torque for a firing angle of 30°.

(b) Input supply power factor.

Solution: Given $I_f = 2$ A

(a) For firing angle 30°

$$V_t = \frac{V_m}{\pi}(1 + \cos \alpha) = \frac{\sqrt{2} \times 230}{\pi}(1 + \cos 30°)$$

$$\boxed{V_t = 193.20 \text{ V}}$$

$$\because \quad V_t = E_b + I_a r_a$$

$$\text{or} \quad V_t = (K \cdot I_f)\,\omega_s + I_a\, r_a$$

$$\text{or} \quad 193.20 = (0.5 \times 2)\frac{2\pi \times 1500}{60} + I_a \cdot 0.8$$

$$\text{or} \quad I_a = \frac{36.12}{0.8} = 45.15 \text{ Amp}$$

$$\text{or} \quad \boxed{I_a = 45.15 \text{ Amp}}$$

Torque $\quad T_e = (K \cdot I_f) \times I_a = (0.5 \times 2) \times (45.15)$

$$\boxed{T_e = 45.15 \text{ N·m}}$$

(b) r.m.s. value of the source current

$$I_{sr} = I_a \left(\frac{\pi - \alpha_1}{2\pi}\right)^{\frac{1}{2}} = 45.15 \left(\frac{180° - 30°}{360°}\right)^{\frac{1}{2}}$$

$$= 45.15 \times 0.645$$

$$\boxed{I_{sr} = 29.14 \text{ Amp}}$$

So, $\quad p.f = \dfrac{V_t \cdot I_a}{V_m \cdot I_{sr}} = \dfrac{193.2 \times 45.15}{\sqrt{2} \times 230 \times 29.14}$

or $\boxed{p.f = 0.92 \text{ lag}}$

Problem 7.5: A 200 V, 100- r.p.m. 10 A separately excited d.c. motor is fed from a single-phase full converter with a.c. source voltage of 230 V, 50 Hz. Armature circuit resistance is 1 Ω. Armature current is continuous. Calculate firing angle for:
(a) Rated motor torque at 500 r.p.m.
(b) Half the rated motor torque at (–500) r.p.m.

Solution: Under rated operating condition

$$V_t = E_a + I_a r_a$$

or $\boxed{V_t = K_m \omega_m + I_a r_a}$

Given, $V_t = 200$ V, $\omega_m = \dfrac{2\pi \times 1000}{60}$

$$I_a = 10 \text{ Amp and } r_a = 1 \ \Omega$$

$\therefore$ $200 = K_m \times \dfrac{2\pi \times 1000}{60} + 10 \times 1$

or $\boxed{K_m = 1.81 \text{ V-s/rad}}$

(a) For rated motor torque, armature current = 10 A
$\therefore$ $V_o = V_t = K_m \omega_m + I_a \cdot r_a$

$$\dfrac{2\sqrt{2} \times 230}{\pi} (\cos \alpha) = 1.81 \times \dfrac{2\pi \times 500}{60} + 10 \times 1$$

or $\dfrac{650.53}{\pi} \cos \alpha = 104.77$

or $\cos \alpha = \dfrac{104.77\pi}{650.53}$

   $\cos \alpha = 0.161 \ \pi$

or $\cos \alpha = 0.505$

$$\boxed{\alpha = 59.61^\circ}$$

(b) For half-rated torque, motor armature current I_a' is equal to

$$I_a' = \frac{1}{2} \times I_a' \text{ (rated)} = \frac{1}{2} \times 10 = 5 \text{ Amp}$$

$\therefore$ $\boxed{I_a' = 5 \text{ Amp}}$

Now, $\dfrac{2\sqrt{2} \times 230}{\pi} \cos \alpha = 1.81 \times \dfrac{2\pi(-500)}{60} + 5 \times 1$

$$= -94.77 + 5$$

$$\frac{2\sqrt{2} \times 230}{\pi} \cos \alpha = -89.77$$

or $\cos \alpha = -0.4335$

or $\boxed{\alpha = 115.68^\circ}$

Problem 7.6: A 200 V, 1500 r.p.m., 10 A, separately excited d.c. motor has an armature resistance of 1 Ω. If is fed from a 1 $-$ ϕ fully-controlled bridge rectifier with an a.c. source voltage of 230 V, 50 Hz. Assuming continuous load current, calculate:
(a) Motor speed at the firing angle of 30° and torque of 15 Nm.
(b) Developed torque at the firing angle of 45° and speed of 1000 r.p.m.

Solution: Under rated operating conditions of the separately-excited d.c. motor.

$$V_t = E_a + I_a r_a$$

$$\boxed{V_t = K_m \omega_m + I_a \cdot r_a}$$

or $200 = K_m \cdot \dfrac{2\pi \times 1500}{60} + 10 \times 1$

$\therefore$ Motor constant, $K_m = \dfrac{200 - 10}{50\pi}$

$$\boxed{K_m = 1.209 \text{ V-S/rad}}$$

(a) For a torque of 15 Nm, motor armature current is given as:

$$I_a = \frac{15}{1.209} = 12.40 \text{ Amp}$$

or $\boxed{I_a = 12.40 \text{ Amp}}$

The equation giving the operation of converter mode is

$$V_o = V_t = E_a + I_a \, r_a$$

or $\dfrac{2\sqrt{2} \times 230}{\pi} \cos 30° = K_m \, \omega_m + 12.40 \times 1$

or $179.33 = 1.209 \, \omega_m + 12.40$

or $\omega_m = \dfrac{166.93}{1.209}$

or $\dfrac{2\pi \times N}{60} = 138.07$

or $\boxed{N = 1318.5 \text{ r.p.m.}}$

∴ $\boxed{\text{Motor speed} = 1318.5 \text{ r.p.m.}}$

(b) For $\alpha = 45°$

$$\dfrac{2\sqrt{2} \times 230}{\pi} \cos 45° = 1.209 \times \dfrac{2\pi \times 1000}{60} + I_a \cdot 1$$

$$146.42 = 126.6 + I_a \cdot 1$$

or $\boxed{I_a = 19.81 \text{ Amp}}$

∴ Motor torque $T_e = K_m I_a$

or $T_e = 1.209 \times 19.81$

$\boxed{T_e = 23.95 \text{ N}_m}$

Problem 7.7: The speed of a separately-excited d.c. motor is controlled by two single-phase full converters, one in the armature circuit and the other in the field circuit. Both converters are fed from the same $1-\phi$, 230 V, 50 Hz source. Armature resistance is 0.5 Ω and field circuit resistance is 200 Ω. Firing angle for field converter is zero and motor constant is 0.8 V-s/A-rad. Armature and field currents are continuous and ripple free. If armature current is 30 A for a firing angle of 45°, then calculate: (a) motor speed, (b) motor torque, (c) input *p.f* of the armature converter, and (d) input *p.f* of the drive.

Solution: For field controlled converter, at $0°$

Firing angle field voltage is given as :

$$V_f = \frac{2V_m}{\pi} = \frac{2 \times \sqrt{2} \times 230}{\pi}$$

or

$$\boxed{V_f = 207.07 \text{ V}}$$

$\therefore$

$$I_f = \frac{V_f}{R_f} = \frac{207.07}{200}$$

or

$$\boxed{I_f = 1.035 \text{ Amp}}$$

(a) For $1 - \phi$ full converter feeding a d.c. motor

$$V_o = V_t = E_a + I_a r_a$$

or at firing angle $45°$

$$\frac{2\sqrt{2} \times 230}{\pi} \cos \alpha = (K \cdot I_f) \frac{2\pi N}{60} + 30 \times 0.5$$

or

$$\frac{2\sqrt{2} \times 230}{\pi} \cos 45° = (0.8 \times 1.035) \frac{2\pi N}{60} + 15$$

or

$$146.422 = 0.0867 N + 15$$
$$131.422 = 0.0867 N$$

or

$$\boxed{N = 1515.1 \text{ r.p.m.}}$$

$\therefore$ motor speed $N = 1515.1$ r.p.m.

(b) Motor torque T_e is given as

$$T_e = (K \cdot I_f) \cdot I_a = (0.8 \times 1.035) \cdot 30 \text{ Nm}$$

$$\boxed{T_e = 24.84 \text{ Nm}}$$

(c) Input power factor of the armature converter is given as,

$$(p.f)_{input} = \frac{V_o}{V_s} \qquad \left[\because I_a = I_{ar} = \frac{146.422}{230} \right]$$

$$\boxed{(p.f)_{input} = 0.6366 \text{ lag}}$$

(d) Input $p.f$ of the drive is given as :

$$= \frac{\text{Power input in Watt}}{\text{Input in } V_A} = \frac{V_t \cdot I_a + V_f \cdot I_f}{V_s \cdot I_a}$$

$$= \frac{146.422 \times 30 + 207.07 \times 1.035}{230 \times 30} = \frac{4606.97}{230 \times 30}$$

Input *p.f* of the drive = 0.6676 lag

Problem 7.8: The speed of a d.c. series motor is controlled by a 3 − φ semiconverter connected to 3 − φ, 400 V, 50 Hz source. The motor constant is 0.4 V-s/A rad. Total field and armature resistance is 1 Ω. Assuming continuous and ripple free armature current at a firing angle of 40° and speed of 1000 r.p.m. determine:
(a) Motor current and motor torque
(b) Power delivered to motor
(c) Reactive power drawn from the supply in VARS.

Solution: For 3 − φ semiconverter feed to a d.c. series motor

$$V_t = E_b + I_a (r_a + r_s)$$

(a) $$\frac{\sqrt{3} \, V_{ml}}{2\pi} (1 + \cos \alpha) = K I_a \, \omega_m + I_a (r_a + r_s)$$

$$\frac{3 \times \sqrt{2} \times 400}{2\pi} (1 + \cos 40°) = 0.4 \, I_a \times \frac{2\pi \times 1000}{60} + I_a \quad (1)$$

or $$477 = 41.88 \, I_a + I_a$$

or $$I_a = \frac{477}{42.88} \text{ Amp}$$

or $$\boxed{I_a = 11.12 \text{ A}}$$

Motor torque

$$T_e = K I_a^2 = 0.4 \times (11.12)^2$$

$$\boxed{T_e = 49.49 \text{ Nm}}$$

(b) Power delivered to motor

$$P = V_t I_a$$

or $$P = 477 \times 11.12$$

$$\boxed{P = 5304.24 \text{ Watt}}$$

(c) Supply line current I_{sr} is given by

$$I_{sr} = I_a \sqrt{\frac{2}{3}} \quad \text{for } \alpha < 60°$$

$$\therefore \text{ Power factor} = \frac{V_t \cdot I_a}{\sqrt{3} \times V_S \cdot I_{Sr}} = \frac{V_t \cdot I_a}{\sqrt{3} \times V_S \cdot I_a \sqrt{\frac{2}{3}}} = \frac{V_t \, I_a}{\sqrt{2} \, V_S \, I_a}$$

$$p.f = \frac{V_t}{\sqrt{2} \, V_S}$$

$$\cos \theta = \frac{477}{\sqrt{2} \times 400} = 0.843$$

$$\therefore \quad \boxed{\theta = 32.51°}$$

$$\therefore \quad \tan \theta = \frac{\text{KVAR}}{\text{kW}}$$

$$\text{or} \quad \tan 32.51° = \frac{\text{KVAR}}{5.304 \, \text{kW}}$$

or Reactive power drawn from the supply in VAR
$$= (\tan 32.51°) \times 5.304 \, \text{KVar} = 3.3813 \, \text{KVar}$$

$$\therefore \quad \boxed{\text{Reactive power} = 3381.3 \, \text{Var}}$$

Problem 7.9: The speed of a 50 kW, 500 V, 120 A, 1500 r.p.m. Separately excited d.c. motor is controlled by a 3 – φ full converter fed from 400 V, 50 Hz supply. Motor armature resistance is 0.1 Ω. Find the range of firing angle required to obtain speeds between 1000 r.p.m. and (–1000) r.p.m. at rated torque.

Solution: $\quad K_m = \dfrac{500 - 12}{2\pi \times \dfrac{1500}{60}} = 3.106$

i.e. Motor constant $K_m = 3.106$

(a) Now, for 1000 r.p.m. at rated load torque
$$V_t = K_m \omega_s + I_d r_a$$

or $\quad \dfrac{3 V_{ml}}{\pi} \cos \alpha = K_m \omega_s + I_d r_a$

or $\quad \dfrac{3 \times \sqrt{2} \times 400}{\pi} \cos \alpha_1 = 3.106 \times \dfrac{2\pi \times 1000}{60} + 120 \times 0.1$

or $\qquad 540.189 \cos \alpha_1 = 325.33 + 12$

or $\qquad\qquad \cos \alpha_1 = \dfrac{337.33}{540.189}$

$\qquad\qquad\qquad \cos \alpha_1 = 0.624$

$\therefore \qquad\qquad\qquad \alpha_1 = \cos^{-1}(0.624)$

or $\qquad\qquad \boxed{\alpha_1 = 51.35^\circ}$

(b) For speed (–1000 r.p.m.) at rated torque

$$\frac{3\sqrt{2} \times 400}{\pi} \cos \alpha_2 = 3.106 \times 2\pi \times \frac{(-1000)}{60} + 120 \times 0.1$$

or $\qquad 540.189 \cos \alpha_2 = -325.33 + 12$

or $\qquad\qquad \cos \alpha_2 = -\dfrac{313.33}{540.189}$

or $\qquad\qquad \cos \alpha_2 = -0.58$

or $\qquad\qquad \boxed{\alpha_2 = -125.45^\circ}$

$\therefore$ Range of firing angle is from

$$\boxed{\alpha_1 = 51.35^\circ \text{ to } \alpha_2 = 125.45^\circ}$$

Problem 7.10: A 100 kW, 500 V, 2000 r.p.m. separately excited d.c. motor is energised from 400 V, 50 Hz 3 – ϕ source through a 3 – ϕ full converter. The voltage drop in conducting thyristor is 1 V. The dc motor parameters are as under

$$r_a = 0.2 \ \Omega, \ K_m = 1.6 \ \text{V-S/rad}, \ L_a = 8\text{mH}$$

Rated armature current = 210 A. No load armature current = 10 % of rated current. Armature current is continuous and ripple free.
(a) Find the no load speed at firing angle of 30°.
(b) Find the firing angle for a speed of 2000 r.p.m. at rated armature current. Also find the supply power factor.
(c) Find the speed regulation for the firing angle obtained in part (b).

Solution: At no load armature current = 10 % of 210 Amp = 21 Amp

(a) The motor terminal voltage

$$V_o = V_t = \frac{3\sqrt{2} \times 400}{\pi} \cos 30^\circ = 467.75 \text{ V}$$

Also, $V_t = E_a + I_a r_a + 1$ V

or $467.75 = K_m \omega_m + 21 \times 0.2 + 1$ V

∴ No-load motor speed is equal to

$$\omega_m = \frac{467.75 - 5.2}{1.6} \text{ rad/sec}$$

$$\omega_m = 289 \text{ rad/sec}$$

∴ $$N_m = \frac{289}{2\pi} \times 60 \text{ r.p.m.}$$

$$\boxed{N_m = 2760.64 \text{ r.p.m.}}$$

(b) At rated armature current and at 2000 r.p.m.

$$V_0 = V_t = K_m \cdot \omega_m + I_a r_a + 1 \text{ V}$$

$$\frac{3\sqrt{2} \times 400}{\pi} \cos \alpha = 1.6 \times \frac{2\pi \times 2000}{60} + 210 \times 0.2 + 1 \text{ V}$$

$$= (335.20 + 42 + 1) \text{ V}$$

or $$\cos \alpha = \frac{378.10 \, \pi}{3\sqrt{2} \times 400}$$

or $\cos \alpha = 0.699$

or $\alpha = \cos^{-1} (0.699)$

or $$\boxed{\alpha = 45.57^\circ}$$

∵ $\alpha < 60^\circ$, so the r.m.s. value of source current is

$$I_{sr} = I_a \sqrt{\frac{2}{3}} = 210 \sqrt{\frac{2}{3}} = 171.46 \text{ A}$$

∴ $$\boxed{I_{sr} = 171.46 \text{ A}}$$

∴ Supply $p.f = \dfrac{V_t \cdot I_a}{\sqrt{3} \, V_S \cdot I_{sr}} = \dfrac{378.1 \times 210}{\sqrt{3} \times 400 \times 171.46}$

$$\boxed{\text{Supply } p.f = 0.669 \text{ lag}}$$

(c) At rated load, speed is 2000 r.p.m., armature terminal voltage $V_t = 378.1$ V and firing angle is 45.57°. At this firing angle, it rated load is reduced to zero, then

$$V_o = V_t = 378.1 = K_m\omega_m + 21 \times 0.2 + 1 \text{ V}$$

or, $$\omega_m = \frac{378.1 - 43}{1.6} \text{ rad/sec}$$

or, $$\omega_m = 209.43 \text{ rad/sec}$$

or, $$N_m = \frac{209.43 \times 60}{2\pi} \text{ r.p.m.}$$

or, $$\boxed{N_m = 1999.9 \text{ r.p.m.}}$$

$$\therefore \text{Speed regulation} = \frac{1999.9 - 2000}{2000} \times 100 \approx 0 \%$$

So, Speed regulation is 0 %.

Problem 7.11: The speed of a separately-excited d.c. motor is controlled by means of two $3 - \phi$ full converters, one in the armature circuit and the other in the field circuit and both are fed from $3 - \phi$. 400 V, 50 Hz supply. Resistance of armature and field circuit are 0.2 Ω and 320 Ω respectively. The motor constant is 0.5 V-S/A.rad. Field converter has zero degree firing angle delay. Armature and field currents have negligible ripple. For rated load torque of 60 Nm at 2000 r.p.m., calculate:
(a) Rated armature current.
(b) Firing angle delay of the armature converter.
(c) Speed regulation at rated load.
(d) Input of the armature converter and the drive at rated load.

Solution: Field voltage V_f is given as

$$V_f = \frac{3\sqrt{2} \times 400}{\pi} \cos 0^\circ$$

$$\boxed{V_f = 540 \text{ V}}$$

$$\therefore \qquad I_f = \frac{V_f}{r_f} = \frac{540}{320}$$

or $$\boxed{I_f = 1.688 \text{ Amp}}$$

(a) Given load torque $T_e = 60 \text{ Nm} = (K \cdot I_f) I_a$
or, $60 \text{ Nm} = (0.5 \times 1.688) I_a$

or $\qquad I_a = \dfrac{60}{0.5 \times 1.688}$

or $\qquad \boxed{I_a = 71.08 \text{ Amp}}$

(b) Let α_1 be the firing angle delay, then

$$\frac{3\sqrt{2} \times 400}{\pi} \cos \alpha_1 = E_b + I_a r_a = (K \cdot I_f)\, \omega_s + I_a r_a$$

$$= (0.5 \times 1.686)\, \frac{2\pi \times 2000}{60} + 71.08 \times 0.2$$

$$= 176.76 + 14.22$$

$$\boxed{540 \cos \alpha_1 = 190.98 \text{ V}}$$

or $\qquad \cos \alpha_1 = \dfrac{190.98}{540}$

or $\qquad \alpha_1 = \cos^{-1} \dfrac{190.98}{540}$

or $\qquad \boxed{\alpha_1 = 69.28^\circ}$

(c) At no load let the speed of motor be ω'_{m_o}

So, $\dfrac{3\sqrt{2} \times 400}{\pi} \cos 69.28^\circ = (0.5 \times 1.688)\, \omega'_m + 0$

or $\qquad \omega'_{m_o} = \dfrac{190.98}{0.5 \times 1.688} = 226.27 \text{ rad/sec}$

In r.p.m. $N'_{m_o} = \dfrac{226.27 \times 60}{2\pi}$ r.p.m. $= 2160.8$ r.p.m.

$\therefore$ Speed regulation $= \dfrac{N'_{m_o} - N_m}{N_m} \times 100 = \dfrac{2160.8 - 2000}{2000} \times 100 = 8.045\ \%$

(d) Input *p.f* of the armature converter is given as :

$$p.f = \frac{V_t \cdot I_a}{\sqrt{3}\, V_S \cdot I_{sr}}$$

$\therefore \qquad I_{sr} = I_a \sqrt{\dfrac{2}{3}}$

$$\therefore \qquad p.f = \frac{V_t \cdot I_o}{\sqrt{2}\, V_s} = \frac{190.98}{\sqrt{2} \times 400}$$

or $\boxed{p.f = 0.3376 \text{ lag}}$

and *p.f* of the drive at rated load is given as :

$$\text{Input } p.f \text{ of the drive} = \frac{\text{Power input in watt}}{\text{Input in } V_A} = \frac{V_t \cdot I_a + V_f \cdot I_f}{V_s \cdot I_a}$$

$$= \frac{190.98 \times 71.10 + 540 \times 1.688}{\sqrt{2} \cdot 400 \times 71.10}$$

$$\boxed{p.f = 0.360 \text{ lag}}$$

Problem 7.12: A d.c. series motor, fed from 400 V d.c. source through a chopper, has the following parameters
$r_a = 0.05\ \Omega$, $r_s = 0.07\ \Omega$, $K = 5 \times 10^{-3}$ Nm/amp^2
The average armature current of 200 A is ripple free. For a chopper duty cycle of 50 % determine : (a) input power from the source, (b) motor speed, and (c) motor torque.

Solution: (a) Input power to the motor

$$= V_t \times I_a = \alpha\, V_S \times I_a = 0.5 \times 400 \times 200 = 40 \text{ kW}$$

(b) For a d.c. series motor

$$V_t = \alpha V_S = E_a + I_a r = K I_a \times \omega_m + I_a r$$

or $0.5 \times 400 = 5 \times 10^{-3} \times 200 \times \dfrac{2\pi N_m}{60} + 200 \times (0.05 + 0.07)$

or $\qquad 200 = 200 \left(5 \times \dfrac{10^{-3} \times 2\pi N_m}{60} + 0.12 \right)$

or $\qquad 60 = 5 \times 10^{-3} \times 2\pi N_m + 7.2$

or $\qquad 52.8 = 5 \times 10^{-3}\ 2\pi N_m$

or $\qquad N_m = \dfrac{52.8}{5 \times 10^{-3} \times 2\pi}$

$$\boxed{N_m = 1680.6 \text{ r.p.m.}}$$

(c) Motor torque
$$T_e = KI_a^2 = 5 \times 10^{-3} \times (200)^2$$

$$\boxed{T_e = 200 \text{ Nm}}$$

Problem 7.13: The chopper used for on-off control of a d.c. separately excited motor has supply voltage of 230 V d.c., an on-time of 10 m sec and off- time of 15 m sec. Neglecting armature inductance and assuming continuous conduction of motor current, calculate the average load current when the motor speed is 1500 r.p.m. and has a voltage constant of $K_v = 0.5$ V/rad per sec. The armature resistance is 3 Ω.

Solution: Duty of chopper is given as

$$\alpha = \frac{T_{on}}{T_{on} + T_{off}} = \frac{10}{10+15} = 0.4$$

$$\therefore \quad \boxed{\alpha = 0.4}$$

Terminal voltage of the motor armature is given as

$$V_t = \alpha\, V_S = E_a + I_a\, r_a$$
$$\alpha\, V_S = K_m \cdot \omega_m + I_a r_a$$

or $0.4 \times 230 = 0.5 \times \dfrac{2\pi \times 1500}{60} + I_a \times 3$

or $92 = 25\pi - 3\, I_a$

or $I_a = \dfrac{92 - 25\pi}{3}$

or $\boxed{I_a = 4.487 \text{ Amp}}$

Problem 7.14: A 230 V d.c. source is connected to a separately-excited d.c. motor through a chopper operating at 500 Hz. The load torque at 1200 r.p.m. is 32.5 Nm. The motor has $r_a = 0$, $L_a = 2$ mH and $K_m = 1.3$ V-S/rad. Motor and chopper losses are neglected.
(a) Calculate the minimum and maximum values of armature current and the armature-current excursion.
(b) Obtain the expressions for armature current during on and off periods of a chopper cycle.

Solution: As the armature resistance is neglected, armature current varies linearly between its minimum and maximum values.

(a) Average armature current

$$I_a = \frac{T_e}{K_m} = \frac{32.5}{1.3} = 25 \text{ Amp}$$

∴

$$\boxed{I_a = 25 \text{ Amp}}$$

Motor e.m.f. is given as

$$E_a = K_m \omega_m = \frac{1.3 \times 2\pi \times 1200}{60}$$

$$\boxed{E_a = 163.36 \text{ V}}$$

Motor input voltage,

$$V_t = \alpha V_S = E_a + I_a r_s = 163.36 + 0$$

∴

$$\alpha = \frac{163.36}{230} = 0.71$$

∴

$$\boxed{\alpha = 0.71}$$

periodic time, $T = \dfrac{1}{f} = \dfrac{1}{500} = 2 \text{ ms}$

on, period $= T_{on} = \alpha T$

$T_{on} = 0.71 \times 2 \text{ ms} = 1.42 \text{ ms}$

off, period $= T_{off} = (1 - \alpha) T$

$= (1 - 0.71) \, 2 \text{ ms}$

$$\boxed{T_{off} = 0.58 \text{ ms}}$$

During on-period T_{on}, armature current will rise which is governed by the equation

$$i_a r_a + L \frac{di_a}{dt} + E_a = V_S$$

or

$$0 + L \frac{di_a}{dt} = V_S - E_a$$

or

$$\frac{di_a}{dt} = \frac{V_S - E_a}{L}$$

$$\frac{di_a}{dt} = \frac{230 - 163.36}{2\,\text{mH}}$$

$$\boxed{\frac{di_a}{dt} = 3332 \text{ A/S}}$$

During off period

$$\boxed{\frac{di_a}{dt} = -\frac{E_a}{L} - 8168 \text{ A/S}}$$

With rising linearly, it is seen that

$$I_{mx} = I_{mn} + \left(\frac{di_a}{dt} \text{ during } T_{\text{on}}\right) \times T_{\text{on}}$$

$$= I_{mn} + 3332 \times 1.42 \times 10^{-3}$$

$$\boxed{I_{mx} = I_{mn} + 4.73} \qquad\qquad \ldots(i)$$

For linear variation between I_{mn} and I_{mx}, average value of armature current

$$I_a = \frac{I_{mx} + I_{nx}}{2} = 25$$

or $I_{mx} + I_{nx} = 50$

or $\boxed{I_{mx} = 50 - I_{nx}}$ $\qquad\qquad \ldots(2)$

From eq. (1) and eq. (2) we get

and $\boxed{\begin{array}{l} I_{nx} = 22.63 \text{ Amp} \\ I_{mx} = 27.36 \text{ Amp} \end{array}}$

$\therefore \qquad \boxed{i_a(t) = 22.63 + 3332t,}$ for $0 \le t \le T_{\text{on}}$

(b) During T_{off}, armature current is given as

$$i_a(t) = I_{mx} + \left(\frac{di_a}{dt}\right)_{T_{\text{off}}} \times t$$

$$\boxed{i_a(t) = 27.36 - 8168t,}$$ for $0 \le t \le T_{\text{off}}$

Problem 7.15: A 220 V, 60 A d.c. series motor, having combined resistance of armature and field of 0.15 Ω is controlled in regenerative braking mode. The d.c. source voltage is 220 V. Motor constant is 0.05 V-s/A.rad. The average motor armature current is rated and ripple free. For a duty cycle of 50% determine:

(a) The power returned to the supply.

(b) Minimum and maximum permissible braking speed.

(c) Speed during regenerative braking.

Solution: (a) Average armature terminal voltage

$$V_t = (1 - \alpha) \, V_S = (1 - 0.5) \times 220$$

or 　　　$V_t = 110$ V

∴ Power returned to the d.c. supply

$$= V_t \cdot I_a = 110 \times 60 = 6.6 \text{ kW}$$

(b) Minimum braking speed is

$$\omega_{mn} = \frac{I_a \cdot r_a}{Km} \text{ rad/sec} = \frac{I_a \cdot r_a}{(K \cdot I_a)} \text{ rad/sed}$$

$$= \frac{0.15}{0.05} \text{ rad/sec} = 3 \text{ rad/sec}$$

∴ 　　$\boxed{\omega_{mn} = 28.65 \text{ r.p.m.}}$

and maximum braking speed is

$$\omega_{mx} = \frac{(V_S + I_a \cdot r_a)}{K_m} = \frac{220 + 60 \times 0.15}{60 \times 0.05} \text{ rad/sec}$$

$$= 76.33 \text{ rad/sec}$$

$\boxed{\omega_{mx} = 728.9 \text{ r.p.m.}}$

(c) When working as a generator during regenerative braking, the generated e.m.f. is

$$E_a = K_m \omega_m = V_t + I_a r_a = 110 + 60 \times 0.15$$

or 　　$\boxed{E_a = 119 \text{ V}}$

∴ Motor speed $\omega_m = \dfrac{119}{0.05 \times 60} \text{ rad/s} = \dfrac{2380}{60} \text{ rad/sec}$

$\boxed{\omega_m = 378.78 \text{ r.p.m.}}$

Problem 7.16: A d.c. chopper is used for regenerative braking of a separately-excited d.c. motor. The d.c. supply voltage is 400 V. The motor has $r_a = 0.2\ \Omega$, $K_m = 1.2$ V-S/rad. The average armature current during regenerative braking is kept constant at 300 A with negligible ripple.

For a duty cycle of 50% for a chopper, determine

(a) Power returned to the d.c. supply
(b) Minimum and maximum permissible braking speed and
(c) Speed during regenerative braking.

Solution: (a) Average armature terminal voltage

$$V_t = (1 - \alpha)\ V_S = (1 - 0.5) \times 400 = 200\ \text{V}$$

Power returned to the d.c. supply

$$= V_t \cdot I_a = 200 \times 300\ \text{W} = 60\ \text{kW}$$

(b) Minimum braking speed is

$$\omega_{mn} = \frac{I_a \cdot r_a}{K_m} = \frac{300 \times 0.2}{1.2}$$

$$\therefore \boxed{\omega_{mn} = 50\ \text{rad/s or } 477.46\ \text{r.p.m.}}$$

Maximum braking speed is

$$\omega_{mx} = \frac{V_S + I_a r_a}{K_m} = \frac{400 + 300 \times 0.2}{1.2}$$

$$\boxed{\omega_{mx} = 383.33\ \text{rad/s or } 3660.06\ \text{r.p.m.}}$$

(c) When working as a generator during regenerative braking, the generated e.m.f. is

$$E_a = K_m \omega_m = V_t + I_a r_a = 200 + 300 \times 0.2 = 260\ \text{V}$$

$$\therefore \text{Motor speed } \omega_m = \frac{260}{1.2}\ \text{rad/sec} = 216.66\ \text{rad/sec}$$

or $$\boxed{\omega_m = 2069\ \text{r.p.m.}}$$

Problem 7.17: (a) For fan type loads, show that rotor current in a 3 − φ induction motor is maximum when slip $S = \dfrac{1}{3}$. State the assumptions made.

(b) A 400 V, 50 Hz, 3 – φ SCIM develops full load torque at 1470 r.p.m. It supply voltage reduces to 340 V, with load torque remaining constant, calculate the motor speed. Assume speed-torque characteristics of the motor to be linear in the stable region. Neglect stator resistance.

Solution: (a) For a fan-type load, torque is proportional to speed square, i.e.

$$T_L \propto \omega_m^2$$

or
$$\boxed{T_L = K \omega_m^2} \qquad \qquad ...(1)$$

Mechanical power developed in motor
$$P_m = (1 - s) P_g$$
As no-load rotational losses are negligible,
$$P_m = \text{Power required by load or}$$
$$(1 - s) P_g = T_L \cdot \omega_m$$

or
$$3 I_2^2 \frac{r_2}{s} (1 - s) = T_L \cdot \omega_m$$

or,
$$I_2 = \left[\frac{T_L \cdot \omega_m \cdot s}{3 r_2 (1 - s)} \right]^{\frac{1}{2}} \qquad \qquad ...(2)$$

$$\because \qquad \omega_m = \omega_s (1 - s), \text{ so putting it in}$$
the above equation we get

$$I_2 = \left[\frac{T_L \cdot \omega_m \cdot s}{3 r_2} \right]^{\frac{1}{2}} = \left[\frac{K \omega_m^2 \cdot \omega_s \cdot s}{3 r_2} \right]^{\frac{1}{2}}$$

$$= \left[\frac{K \omega_s^3 (1 - s)^2 \cdot s}{3 r_2} \right]^{\frac{1}{2}} \quad I_2 = \sqrt{s} \cdot (1 - s) \left[\frac{K \cdot \omega_s^3}{3 r_2} \right]^{\frac{1}{2}}$$

The slip at which rotor current I_2 becomes maximum can be obtained by differenciating I_2 w.r.t. slips. So

$$\frac{dI_2}{ds} = \frac{1}{2} \cdot \frac{1}{\sqrt{3}} (1 - s) \left[\frac{K \omega_s^3}{3 r_2} \right]^{\frac{1}{2}} + \sqrt{s} \ (s - 1) \left[\frac{K \omega_s^3}{3 r_2} \right]^{\frac{1}{2}} = 0$$

or $\quad \dfrac{1-s}{2} = s$

or $\quad \boxed{s = \dfrac{1}{3}}$

This shows that for a fan type load I_2 is maximum at a slip of $\dfrac{1}{3}$.

(b) $\qquad \omega_r = 1470$ r.p.m.

$\therefore \qquad \text{Slip} = \dfrac{1500 - 1470}{1500}$

$\qquad\qquad s = 0.02$

When slip s is small, become quite large as compare to x. So torque equation is given as

$$T_e = \frac{3V^2}{\omega_s} \cdot \frac{s}{r_2}$$

$\therefore$ For the same load torque

$$\frac{T_{e_1}}{T_{e_2}} = \frac{V_1^2 s_1}{V_2^2 s_2} = 1$$

or $\qquad \dfrac{V_1^2}{V_2^2} = \dfrac{s_2}{s_1}$

or $\qquad s_2 = \dfrac{V_1^2}{V_2^2} \times s_1$

or $\qquad s_2 = \dfrac{(400)^2}{(340)^2} \times 0.02 = 0.02768$

$\therefore$ Motor speed $= 1500 \,(1 - s_2) = 1500 \,(1 - 0.02768)$

$\qquad\qquad\qquad = 1458.4$ r.p.m.

Problem 4.18: A $3 - \phi$, 400 V, 20 kW, 970 r.p.m., 50 Hz, delta connected Induction motor has rotor leakage impedance of $0.5 + J\,2.00\ \Omega$. Stator leakage impedance and rotational losses are assumed negligible. If this motor is energised from a source of $3 - \phi$, 400 V, 90 Hz, then compute:

(a) The motor speed at rated torque.
(b) The slip at which maximum torque occurs.
(c) The maximum torque.

Solution: Full-load torque $(T_e)_{fl} = \dfrac{P}{\omega_m}$

or $\qquad (T_e)_{fl} = \dfrac{20000}{2\pi \times 970} \times 60$

$\qquad (T_e)_{fl} = 196.89$ N.m

(a) At 90 Hz, synchronous speed,

$$\omega_s = \frac{4\,\pi f_1}{p} = \frac{4\,\pi \times 90}{4}$$

or $\qquad \boxed{\omega_s = 90\,\pi \text{ rad/sec}}$

Rotor impedance at 90 Hz

$$= 0.5 + J2 \times \frac{90}{50} = 0.5 + J3.6 \approx J3.6 \ \Omega$$

$\therefore \qquad T_{efl} = 196.89 = \dfrac{3}{90\pi} \dfrac{(400)^2}{(3.6)^2} \times \dfrac{0.5}{S_{fl}}$

$\therefore \qquad$ Full load slip $S_{fl} = 0.332$

$\therefore \qquad$ Motor speed $= \dfrac{120 \times 90}{4} \ (1 - 0.332)$

$\qquad \boxed{\text{Motor speed} = 1819 \text{ r.p.m.}}$

(b) Slip at which maximum torque occure

or $\qquad S_m = \dfrac{r_2}{x_2} = \dfrac{0.5}{3.6}$

$\qquad \boxed{S_{mn} = 0.138}$

(c) Maximum torque

$$T_{e,\,max} = \frac{3}{\omega_s} \frac{V_1^2}{2x_2} = \frac{3 \times (400)^2}{90\pi \times 2 \times 3.6}$$

$\qquad \boxed{T_{e,\,max} = 235.785 \text{ Nm}}$

Problem 7.19: A 3 – ϕ, 500 V, 20 kW, 1440 r.p.m., 50 Hz, star-connected Induction motor has rotor leakage impedance of 0.4 + J 1.6 Ω. Stator leakage impedance and rotational losses are assumed negligible. If this motor is energised from 90 Hz, 500 V, 3 – ϕ source, then calculate:

(a) The motor speed at rated load
(b) The slip at which maximum torque occure
(c) The maximum torque.

Solution: Full load torque

$$(T_e)_{fl} = \frac{P}{\omega_m} = \frac{20 \times 1000}{2\pi \times \dfrac{1440}{60}}$$

$$\boxed{(T_e)_{fl} = 132.62 \text{ Nm}}$$

(a) At 90 Hz, synchronous speed

$$\omega_s = \frac{4\pi + f_1}{P} = \frac{4\pi \times 90}{4}$$

or $$\boxed{\omega_s = 90 \pi \text{ rad/sec}}$$

Rotor impedance at 90 Hz

$$= 0.4 + J\, 1.6 \times \frac{90}{50} = 0.4 + J\, 2.88 \approx J\, 2.88 \ \Omega$$

$$\therefore \qquad (T_e)_{fl} = \frac{3}{90\pi} \frac{\left(\dfrac{500}{\sqrt{3}}\right)^2}{(2.88)^2} \times \frac{0.4}{S_{fl}} = 132.62$$

$\therefore$ Full load slip

$$\boxed{S_{fl} = 0.321}$$

$\therefore$ Motor speed $= \dfrac{120 \times 90}{4} (1 - 0.321) = 1831.88$ r.p.m.

(b) Slip at which maximum torque occurs is given by

$$S_{\max} = \frac{0.4}{2.88} = 0.138$$

$$\therefore \quad \boxed{S_{max} = 0.318}$$

(c) Maximum torque $T_{e.max} = \dfrac{3}{\omega_s} \dfrac{V_1^2}{2 x_2} = \dfrac{3}{90\pi} \dfrac{\left(\dfrac{500}{\sqrt{3}}\right)^2}{2 \times 2.88}$

$$\boxed{(T_e)_{max} = 153.50 \text{ Nm}}$$

Problem 7.20: A $3 - \phi$, 15 kW, 420 V, 4 pole, 50 Hz delta connected induction motor has the following per-phase parameter referred to stator

$$r_1 = 0.5 \ \Omega, \quad r_2 = 0.4 \ \Omega, \quad x_1 = x_2 = 1.5 \ \Omega, \quad x_m = 0$$

If this motor is operated at 210 V, 25 Hz with DoL starting; calculate:
(a) Current and *p.f* at the instant of starting and maximum torque conditions; compare the results with normal values.
(b) Starting and maximum torque and compare with normal values.

Solution: Let I_n be the normal operating current and I_d be the reduced-voltage, reduced freq current

(a) Starting current

$$I_n = \frac{V_1}{\left[\left(r_1 + \dfrac{r_2}{s}\right)^2 + (x_1 + x_2)^2 \right]^{\frac{1}{2}}}$$

$$= \frac{420}{\left[\left(0.5 + \dfrac{0.4}{1}\right)^2 + (1.5 + 1.5)^2 \right]^{\frac{1}{2}}} = \frac{420}{3.132} = 134.09 \text{ Amp}$$

$$\therefore \quad I_n = 134.09 \text{ Amp}$$

$$I_d = \frac{210}{[3.060]^{\frac{1}{2}}} \quad \left[\because (x_1 + x_2) \frac{25}{50} \right]$$

$$= \frac{210}{1.74} = 120.04$$

$$\boxed{I_d = 120.04 \text{ Amp}}$$

$$(\cos \theta)_n = \frac{0.9}{(3^2 + 0.9^2)^{\frac{1}{2}}} = 0.287$$

$$(\cos \theta)_d = \frac{0.9}{(1.5^2 + 0.9^2)^{\frac{1}{2}}} = 0.514$$

(b) Starting torque

$$T_{en} = \frac{3}{\omega_{sn}} I_n^2 \cdot \frac{r_2}{s} = \frac{3}{50\pi} (134.09)^2 \times \frac{0.4}{1}$$

$$\boxed{T_{en} = 137.38 \text{ Nm}}$$

$$T_{ed} = \frac{3}{\omega_{sd}} I_d^2 \times \frac{r_2}{s} = \frac{3}{25\pi} (120.04)^2 \times \frac{0.4}{1}$$

$$\boxed{T_{ed} = 220.16 \text{ Nm}}$$

Maximum torque : The slip at which maximum torque occurs is given by

$$\frac{0.4}{S_{m.n}} = \sqrt{(0.5)^2 + 3^2}$$

or

$$\boxed{S_{m.n} = 0.131}$$

$$\therefore \quad I_n = \frac{420}{\left(\left(0.5 + \frac{0.4}{0.131}\right)^2 + 3^2\right)^{\frac{1}{2}}} = 90.486 \text{ Amp}$$

$$T_{e.m.n.} = \frac{3}{50\pi} (90.486)^2 \times \frac{0.4}{0.131}$$

$$\boxed{T_{e.m.n.} = 475.66 \text{ Nm}}$$

At reduced voltage and reduced frequency

$$S_{m.d} = \frac{0.4}{\left[0.5^2 + \left(\frac{3}{2}\right)^2\right]^{\frac{1}{2}}} = 0.252$$

$$I_d = \cfrac{210}{\left(\left(0.5 + \cfrac{0.4}{0.252}\right)^2 + (1.5)^2\right)^{\frac{1}{2}}}$$

$$\boxed{I_d = 81.86 \text{ Amp}}$$

$$T_{e.m.d} = \frac{3}{25\pi}(81.86)^2 \times \frac{0.4}{0.252} = 404.702 \text{ Nm}$$

It is seen from above that at reduced voltage and at reduced frequency:

(1) Current gets reduceses where as $p.f$ is improved at the starting.

(2) The maximum torque, however, get reduced.

Problem 7.21: A 420 V, 6 pole, 50 Hz, 3 – ϕ, star connected IM has $r_1 = 0$, $x_1 = x_2 = 1.2$ Ω, $r_2 = 0.5$ Ω and $x_m = 50$ Ω as its per-phase parameters referred to stator. This IM is fed from (i) constant-voltage source of 242.5 V per phase and (ii) constant-current source 30 A.

For both types of sources (i), and (ii) calculate:

(a) The slip for maximum torque.

(b) The starting and maximum torques.

(c) The supply voltage required to sustain the constant current at the maximum torque.

Solution: The equivalent circuit for this IM is drawn as :

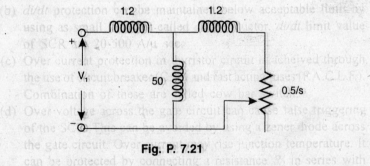

Fig. P. 7.21

Its Thevenin's equivalent circuit is drawn as :

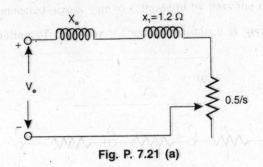

Fig. P. 7.21 (a)

Where $\quad x_e = \dfrac{1.2 \times 50}{51.2} = 1.17 \ \Omega$

and $\quad V_e = 242.5 \times \dfrac{50}{51.2} = 236.8 \ \text{V}$

(a) (i) It is seen from the above equivalent circuit that the slip at which maximum torque occurs is given as :

$$S_m = \frac{0.5}{2.37} = 0.2109$$

(ii) For constant current operation of IM, slip S_m is given by

$$S_{mt} = \frac{0.5}{50 + 1.2} = 0.00976$$

(b) (i) Synchronous speed $\omega_s = \dfrac{4\pi f_1}{P} = \dfrac{4\pi \times 50}{6}$ rad/s

$$= 104.71 \ \text{rad/sec}$$

for constant voltage input, $T_{e.st}$ is given as

$$T_{e.st} = \frac{3}{\omega_s} \cdot (I_{2st})^2 \cdot \frac{r_2/(0.81)^2}{1}$$

$$= \frac{3}{50\pi} \frac{(236.8)^2}{(0.5)^2 + (2.37)^2} \times 0.5$$

$$= 91.26 \ \text{Nm} \times 1.5 = 136.71 \ \text{Nm}$$

$$T_{e.m} = \frac{3}{w_s} \frac{V_e^2}{2(x_2 + x_e)}$$

$$= \frac{3}{50\pi} \times \frac{(236.8)^2}{2 \times (2.37)(0.81)^2} = 225.93 \text{ Nm} \times 1.5$$

$$\boxed{T_{e.m} = 338.93 \text{ Nm}}$$

(ii) For constant current input, $T_{e.sf}$ is

$$T_{e.sf} = \frac{3}{50\pi} \frac{(30 \times 50)^2}{0.5^2 + 51.2^2} \times \frac{0.5}{(0.81)^2} = 12.19 \text{ Nm}$$

$$\boxed{T_{e.sf} = 12.19 \text{ Nm}}$$

Maximum torque is

$$T_{e.m} = \frac{3}{50\pi} \frac{(30 \times 50)^2}{2 \times 51.2 \,(0.81)^2}$$

$$\boxed{T_{e.m} = 629.64 \text{ Nm}}$$

(c) At maximum torque, $S_{mT} = 0.00976$, the magnetizing current I_m is given as :

$$I_m = I_1 \cdot \frac{\dfrac{r_2}{s} + Jx_2}{\dfrac{r_2}{s} + J\,(x_2 + x_m)} = 30 \times \frac{\dfrac{0.5}{0.00976} + J1.2}{\dfrac{0.5}{0.00976} + J(1.2 + 50)}$$

$$= 30 \; \frac{(51.22 + J1.2)}{(51.22 + J(51.2))} = 30 \; \frac{(51.23 \angle 1.34°)}{(72.42 \angle 45°)}$$

$$= 30 \times 0.707 \; \angle 43.64° = 21.22 \; \angle 43.64°$$

∴ Supply voltage required to sustain constant current of 30 A is given by

$$V_1 = \sqrt{3} \times (21.22) \times 50$$

$$\boxed{V_1 = 1837.88 \text{ V}}$$

Reader's Notes

Reader's Notes

Reader's Notes

Reader's Notes